walkermaths

NUMERACY
PRACTICE

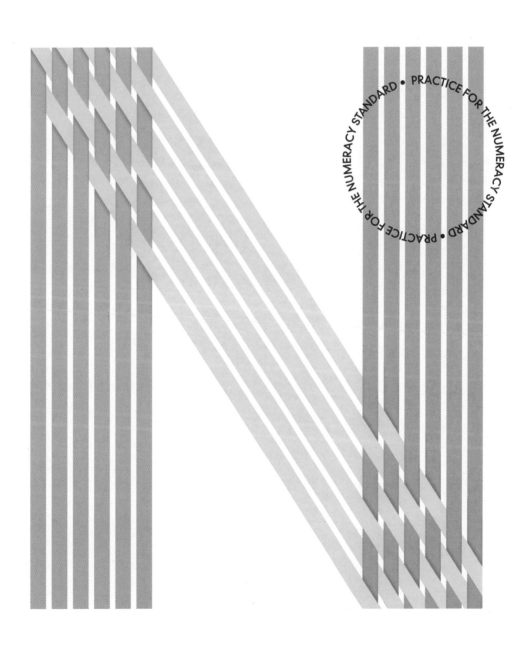

PRACTICE FOR THE NUMERACY STANDARD • PRACTICE FOR THE NUMERACY STANDARD •

Victoria and Charlotte Walker

Walker Maths: Numeracy Practice
1st Edition
Charlotte Walker
Victoria Walker

Cover design: Cheryl Smith, Macarn Design
Text design: Cheryl Smith, Macarn Design
Production controller: Siew Han Ong

Any URLs contained in this publication were checked for currency during the production process. Note, however, that the publisher cannot vouch for the ongoing currency of URLs.

For product information and technology assistance,
in Australia call **1300 790 853**;
in New Zealand call **0800 449 725**

For permission to use material from this text or product, please email
aust.permissions@cengage.com

National Library of New Zealand Cataloguing-in-Publication Data
A catalogue record for this book is available from the National Library of New Zealand

978 01 7047447 4

Cengage Learning Australia
Level 5, 80 Dorcas Street
Southbank VIC 3006 Australia

Cengage Learning New Zealand
For learning solutions, visit **cengage.co.nz**

Printed in China by 1010 Printing International Limited.
9 26 25 24

CONTENTS

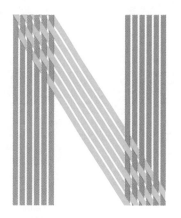

Glossary 4

Process ideas 6

1 Formulate approaches to solving
  problems 6

2 Use mathematics and statistics 8

3 Explain the reasonableness of
  responses 10

Practice 1 16

Practice 2 18

Practice 3 20

Practice 4 22

Practice 5 24

Practice 6 26

Practice 7 28

Practice 8 30

Practice 9 32

Practice 10 34

Practice 11 36

Practice 12 38

Practice 13 40

Practice 14 42

Practice 15 44

Practice 16 46

Practice 17 48

Practice 18 50

Practice 19 52

Practice 20 54

Practice 21 56

Practice 22 58

Practice 23 60

Practice 24 62

Practice 25 64

Practice 26 66

Practice 27 68

Practice 28 70

Practice 29 72

Practice 30 74

Practice 31 76

Practice 32 78

Practice 33 80

Practice 34 82

Practice 35 84

Practice 36 86

Practice 37 88

Practice 38 90

Practice 39 92

Practice 40 94

Answers 96

Term	Meaning/Picture/Example
Mean	
Median	
Mode	
Average	
Range	
Reflect	
Rotate	
Translate	
Enlarge	
Scale factor	
Order of rotational symmetry	
Order of line symmetry	
Clockwise/ anticlockwise	
Horizontal/vertical	
Ratio	

Term	Meaning/Picture/Example
Less than (<)	
More than (>)	
Century	
1 million using digits/ numerals	
1 billion using digits/ numerals	
Credit	
Debit	
Balance	
Simple interest	
Time and a half	
GST	
Surface area	
Dimensions	
Hectares (ha)	
Odometer	
Sector	
Capacity	

ISBN: 9780170474474

Process ideas

- In the Numeracy assessment you will need to be able to solve mathematical problems in a range of meaningful situations using three process ideas.

You will need to:

1 **Formulate** approaches to solving problems. This means working out how to solve a problem.
2 **Use** mathematics and statistics in a range of situations.
3 **Explain** whether answers and statements are reasonable.

1 Formulate approaches to solving problems

- These questions will require you to work out **how** to solve the problem.
- Usually there are several steps needed to find the answer.

Example:
Amanda's letterbox is 900 m from her house. It takes her 12 minutes to walk to it.
Explain how you could calculate her speed in km/hour.
You could do this in two ways:

i Show the calculation with speech bubbles:

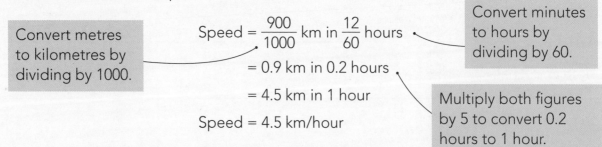

Convert metres to kilometres by dividing by 1000.

$$\text{Speed} = \frac{900}{1000} \text{ km in } \frac{12}{60} \text{ hours}$$

Convert minutes to hours by dividing by 60.

$$= 0.9 \text{ km in } 0.2 \text{ hours}$$

$$= 4.5 \text{ km in 1 hour}$$

Multiply both figures by 5 to convert 0.2 hours to 1 hour.

$$\text{Speed} = 4.5 \text{ km/hour}$$

ii Explain each stage of the calculation in sentences:
1 kilometre = 1000 m so divide 900 m by 1000 to convert it to kilometres: 0.9 km
1 hour = 60 minutes so divide 12 minutes by 60 to convert it to hours: 0.2 hours
To convert 0.9 km in 0.2 hours to distance walked in 1 hour, multiply both values by 5.
Her speed was 4.5 km/hour.

Answer the following questions.

1 Josephine went shopping for four house plants. Single plants cost $12.95.
However, the shop has two deals:
Deal 1: Two plants for $22.50.
Deal 2: 20% off if you buy three plants or more.
Which is the best deal for four plants?

_____

_____

ISBN: 9780170474474

2 Miru visited a cave where a stalactite is growing at an average rate of 0.02 mm per month.
She calculated that, to the nearest year, it would take 4167 years for it grow a metre. Explain how she could have reached this conclusion.

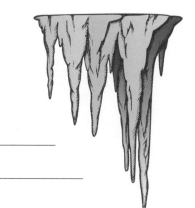

_____

_____

_____

3 Sarah celebrated Matariki by baking some cookies.
These quantities of ingredients will make 30 cookies:

225 g butter
225 g caster sugar
1 egg
450 g flour
1 tsp baking powder
2 tsp vanilla essence

How much butter will she need to make 72 cookies?   _____ g

4 It takes a cruise ship four and a half days to cover the 2160 km from Auckland to Fiji. Explain how you could calculate the cruise ship's average speed in kilometres per hour.

_____

_____

_____

5 The graph shows the cost of hiring a small digger for up to five days.

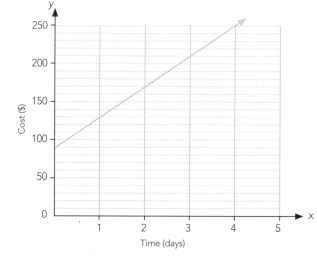

Complete the following statement:

The cost of hiring the digger is $_____

plus $_____ for each day.

* In these questions you will be told what to calculate.

Example:
Rena wants to ring her friend in Perth, Australia. They are four hours behind New Zealand. If Rena rings them at 19:42 New Zealand time, what is the time in Perth?

☐ 5.42 pm   ☐ 3.42 am   ☑ 3.42 pm   ☐ 13:42

19:42 is 7.42 pm in 12-hour time. Four hours earlier it is 3.42 pm.

Answer the following questions.

1 a   Four tables with trapezium-shaped tops are arranged to make a square.

x

50 mm

675 mm

1750 mm

Calculate the size of angle *x*.   _____ °

b   Which calculation would give you the total shaded area?

☐ $1750^2 - 50^2$ ☐ $1750^2 + 50^2$ ☐ $4 \times 1750 \times 675$ ☐ $4 \times \dfrac{1750 \times 675}{2}$

2   Fabian's monthly activity goal is to complete 155 000 steps. If the month is 31 days long, how many steps on average does he need to take each day?

_____ steps

3 **a** The giant moa is an extinct New Zealand flightless bird.

Estimate the height of this moa.

_____ m

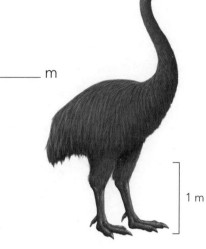

1 m

b It is believed that once there were 2.5 million of the giant moa.
Write this number using digits only.

_____

c Each egg laid by a female moa has a mass about 2% of her body mass.

If the mass of a female moa was 230 kg, estimate the mass of
one egg.

_____ kg

d It is thought that moa could travel at 112 km/hour. Assuming it could keep this
pace up, how long (to the nearest minute) would it have taken to run 84 km, the
distance from Ashburton to Christchurch?

_____ minutes

4 The prices charged for the use of electric scooters by two companies are graphed
below.

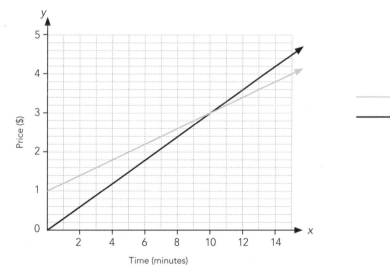

——— Blue Moon scooters
——— Black Cat scooters

a How much would a 14-minute trip with Blue Moon scooters
cost you?

$_____

b What is the cost per minute to hire a scooter from Black
Cat scooters?

_____ cents

3 Explain the reasonableness of responses

- In these questions you will be given a statement and you need to state whether you agree, disagree or can't tell for sure. Then you **must** explain your choice.
- Sometimes more than one of the three options could be correct, depending on your explanation.
- When answering these questions, where possible use numbers to support your answer.
- When answering questions about calculations, use words in your explanations.
- You **must** base your answer on information given to you, **not** you own experience.

Example:

The graph shows the results for the Paradise High School boys' hockey team over the last three years.

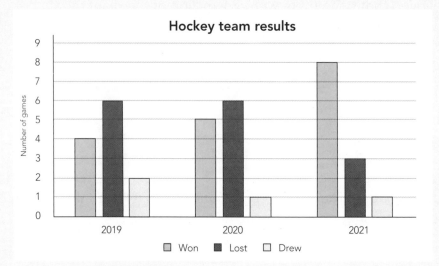

Max says that the team will almost certainly win most of their games in 2022.

☑ Agree   ☐ Disagree   ☐ Can't tell for sure

Explain your answer.

The proportion of games that the team has won has increased each year from $\frac{4}{12}$ in 2019 to $\frac{8}{12}$ in 2021, so it is very likely that they will win most of their games in 2022.

Where possible, use numbers to strengthen your argument.

or

☐ Agree   ☐ Disagree   ☑ Can't tell for sure

Explain your answer.

Although the proportion of games that the team has won has increased each year, it may be that most of their strongest players or the coach left school at the end of 2021, so we cannot tell for certain.

Answer the following questions.

1 Here is a copy of Jerry's bank account statement:

	Credit means how much money is put into his account.	Debit means how much money is taken out of his account.	Balance means how much money is left in his account.

Date	Description	Credits	Debits	Balance
Latest → 11 Aug 22	Bike shop		$654.90	$2.63
11 Aug 22	Power		$206.44	$657.53
11 Aug 22	Rent		$250.00	$863.97
11 Aug 22	Pay	$1093.86		$1133.97
10 Aug 22	Supermarket		$63.53	$20.11
Oldest → 09 Aug 22				$83.64

Jerry should be worried that he is spending more than he earns.

☐ Agree   ☐ Disagree   ☐ Can't tell for sure

Explain your answer.

_____

_____

2 Ria went shopping and put these items in her basket:

 **$36**    **$79.95**    **$12.90**

Ria uses this calculation to estimate how much she will pay.

$$\$40 + \$80 + \$15 = \$135$$

Ria's method is a good estimate of the cost.

☐ Agree   ☐ Disagree   ☐ Can't tell for sure

Explain your answer.

_____

_____

3 The city council announced that compared with 2021, the number of people using buses had doubled in 2022. They used this graph to make their point.

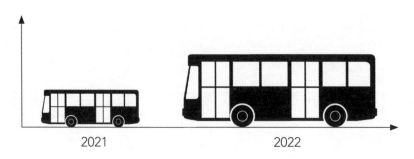

2021   2022

This graph is a fair way to show that the number of people using buses had doubled.

☐ Agree   ☐ Disagree   ☐ Can't tell for sure

Explain your answer.

_____

_____

4 When someone has an accident they are sometimes supported by the Accident Compensation Corporation (ACC).
The graph shows the numbers of ACC claims for rugby injuries in 2021.

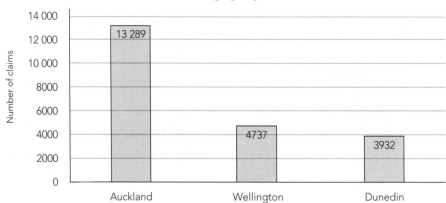

Rugby injuries

Findlay says rugby players are more likely to make an ACC claim for a rugby injury in Auckland than in Dunedin.

☐ Agree   ☐ Disagree   ☐ Can't tell for sure

Explain your answer.

_____

_____

5 Admission to the Fun Park is $28 for adults and $15 for children. However it advertises this deal:

> ## GREAT DEAL!   THIS WEEK ONLY!
> ### $80 FAMILY PASS: 2 ADULTS AND UP TO 3 CHILDREN!

The Fun Park really is offering a great deal for families.

☐ Agree   ☐ Disagree   ☐ Can't tell for sure

Explain your answer.

_____

_____

---

6 Hillary sells jam, chutney and honey at her local Saturday market. The graph shows the number of jars of each product that she has sold over the last three weeks.

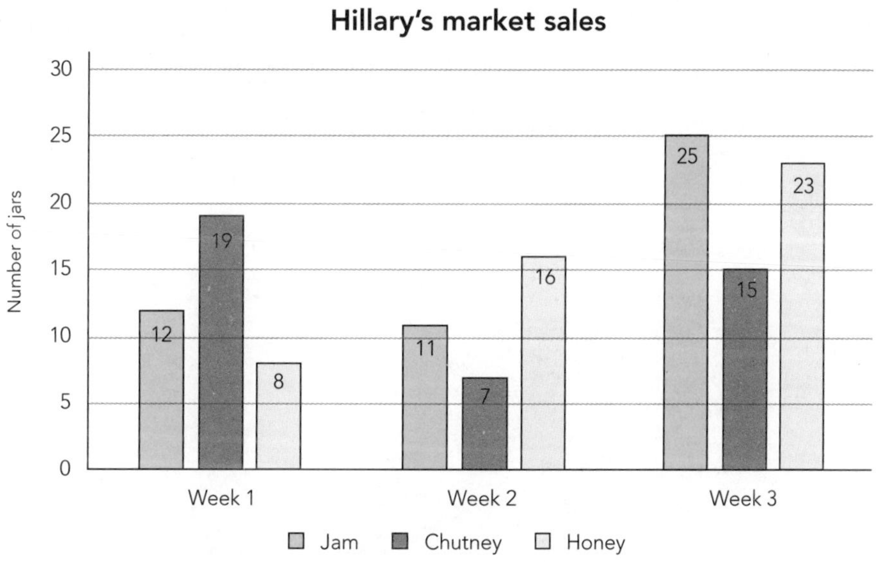

Hillary's market sales

Bar chart showing Number of jars (y-axis 0 to 30) across Week 1, Week 2, Week 3. Jam, Chutney, Honey. Week 1: Jam 12, Chutney 19, Honey 8. Week 2: Jam 11, Chutney 7, Honey 16. Week 3: Jam 25, Chutney 15, Honey 23.

☐ Jam   ☐ Chutney   ☐ Honey

Her brother tells her that she is bound to sell at least 20 jars of honey next week.
Do you agree with her brother? Explain your answer using the information on the graph.

☐ Agree   ☐ Disagree   ☐ Can't tell for sure

Explain your answer.

_____

_____

7 Henare is given a challenge. He must draw the shortest line possible connecting the four corners of the rectangle without taking his pen off the page.
Four routes are shown on the right.

Ben says the N-shaped route is the shortest.

☐ Agree   ☐ Disagree   ☐ Can't tell for sure

Explain your answer.

_____

_____

8 On Tuesday the principal surveyed 187 Year 10 students anonymously to find out how much time each had spent on homework the previous night.

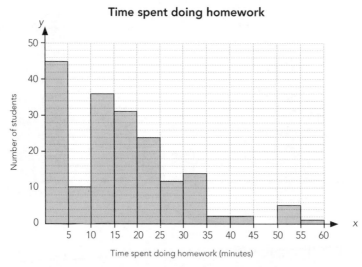

Time spent doing homework

His conclusion was that over half of students do less than 15 minutes' homework each night.

☐ Agree   ☐ Disagree   ☐ Can't tell for sure

Explain your answer.

_____

_____

9 Pukeko nests are unusual because they normally contain eggs from several females. The number of eggs in 10 nests were counted. These are the results:

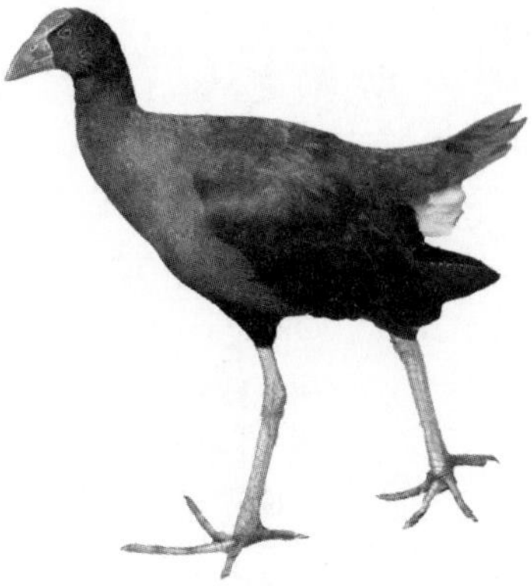

2   12   13   15   16   16   17   17   17   18

Anna said that the mean gives the best measure of the centre of the data.

☐ Agree   ☐ Disagree   ☐ Can't tell for sure

Explain your answer.

_____

_____

10 Each student in the school voted for which of the native birds found in the school grounds was their favourite. The results are shown on the pie graph.

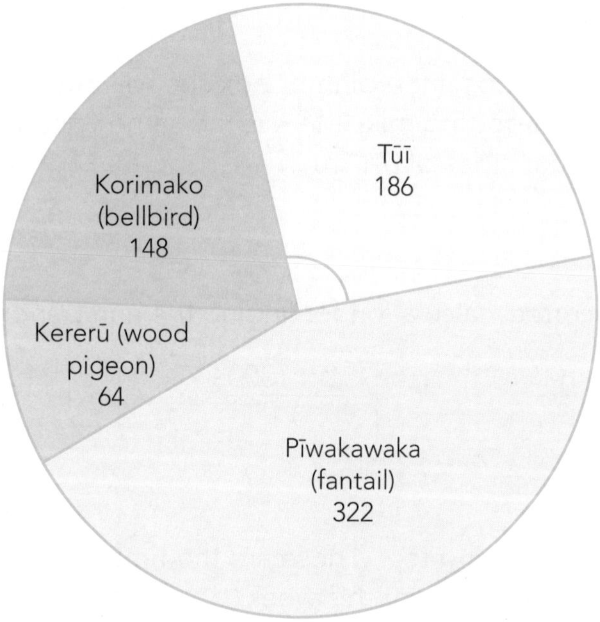

Tūī
186

Korimako
(bellbird)
148

Kererū (wood
pigeon)
64

Pīwakawaka
(fantail)
322

Heidi said the angle at the centre of the sector for tūī was 90°.

☐ Agree   ☐ Disagree   ☐ Can't tell for sure

Explain your answer.

_____

_____

Practice 1

1　**a**　At a garden store, 6 L of potting mix costs $5.50, 12 litres cost $11.00. Which graph shows the relationship between the amount of potting mix and its cost?

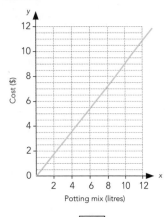

A ☐

B ☐

C ☐

b　The internal dimensions of Ariana's planter box are 2.4 m long by 1.2 m and it is 0.8 m deep. Calculate its volume.

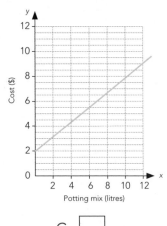

Volume = _____ m³

c　She estimates that to fill the planter box she will need 2 cubic metres of potting mix. 1 litre is the same volume as a cube that has sides of 0.1 m.

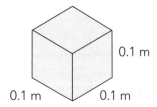

0.1 m

0.1 m　0.1 m

Which is the correct calculation for finding the number of litres in 2 cubic metres?

☐ $\dfrac{2 \times 100}{0.1^3}$　　☐ $\dfrac{2 \times 1000}{0.1^3}$　　☐ $\dfrac{2}{0.1^3}$　　☐ 2×0.1^3

2　Joe is studying for his driving test. These are the times he spent studying on each day of this week:

Day	Time
Monday	27 minutes
Tuesday	48 minutes
Wednesday	35 minutes
Thursday	1 hour
Friday	22 minutes

a　What is the mean of the times he spent studying each day?

_____ minutes

b　On Saturday, he studied for 72 minutes without a break. If he started at 10.17 am, at what time did he finish?

☐ 10.19 am　　☐ 10.29 am　　☐ 11.19 am　　☐ 11.29 am

3 Jonte has made this shape from some blocks.

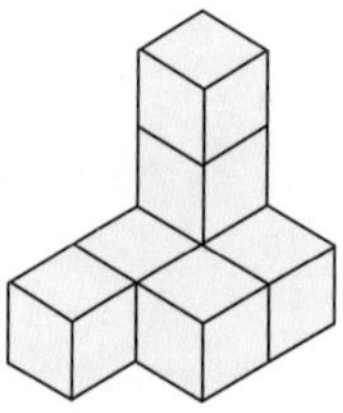

a Which diagram below shows the top view of this solid?

☐ 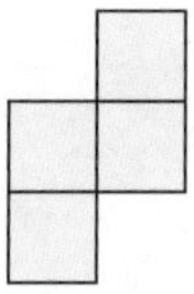 ☐ 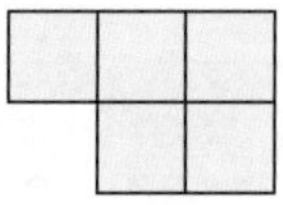 ☐ 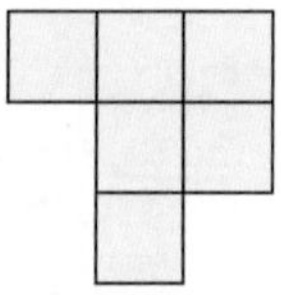 ☐ 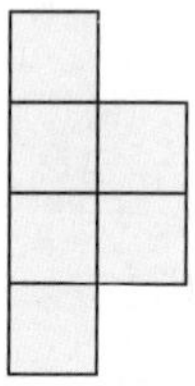

b Jonte wants to add to his shape and make a cube that is three blocks high, three blocks deep and three blocks long.

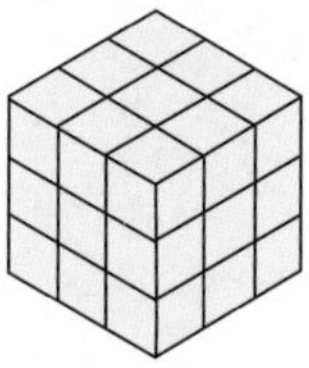

How many **more** blocks will Jonte need? ________

c All the blocks making up the cube were stuck together. Then the completed cube was dipped in green paint. How many faces of the blocks will be green?

☐ 27 ☐ 30

☐ 36 ☐ 54

4 **a** This pie graph shows students' favourite sports.

Favourite sports

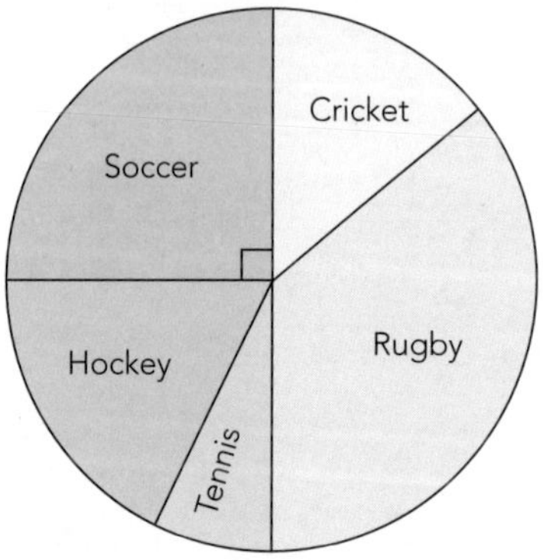

Place a tick beside any true statements (there may be more than one):

☐ Cricket was more popular than hockey.

☐ One quarter of the students favoured soccer.

☐ Soccer was less popular than hockey.

☐ Most students voted for rugby or soccer.

b The angle at the centre of the pie graph for the hockey sector is 60°.

If 252 students answered this survey, how many preferred hockey? ____________

c Rugby is the favourite sport for the largest number of students.

☐ Agree ☐ Disagree ☐ Can't tell for sure

Explain your answer.

____

____

ISBN: 9780170474474

Practice 2

1 **a** Hana and her friends are tramping on Stewart Island. They are going to tramp from Mason Bay Hut to Freshwater hut.

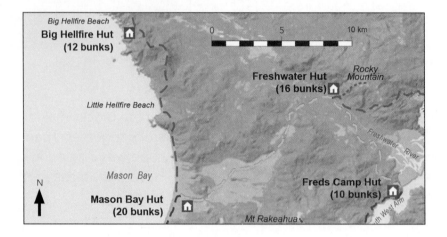

  a Use the scale on the map to estimate how many kilometres
  they will need to tramp.   _____ km

  b In what general direction will they be tramping?   _____

  c The distance from Freshwater Hut to Freds Camp Hut is 11 km. They estimate
  that they can tramp at 2.5 km/hour.

  At that rate, how long will it take them to walk this section of the track?
  Give your answer in hours and minutes.

                                      _____ hours and _____ minutes

2 Isabelle spun this spinner 60 times.

  Which table is most likely to show the results?

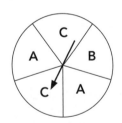

☐   ☐   ☐   ☐

Letter spun	Number of spins
A	30
B	29
C	31

Letter spun	Number of spins
A	11
B	9
C	40

Letter spun	Number of spins
A	25
B	12
C	23

Letter spun	Number of spins
A	24
B	8
C	28

3   A piece of A3 paper is twice the size of a piece of A4 paper. The dimensions of an A3 piece of paper are 297 by 420 mm.

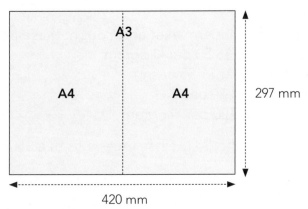

What is the area of a piece of **A4** paper?

☐ 124 740 mm²   ☐ 31 185 mm²   ☐ 1434 mm²   ☐ 62 370 mm²

4   **a**   What is the most accurate way of estimating the total cost of these four objects?

$82   **$53**   **$98**   **$49**

☐ 80 + 50 + 90 + 50   ☐ 80 + 50 x 2 + 90

☐ 80 + 50 x 2 + 100   ☐ 80 + 50 + 100 + 40

b   If the price of the handbag is reduced by 15%, calculate the reduced price.   _____

c   A sign above the hoodie says 'Buy one, get one half price'. Nathan and Sam decide to buy two hoodies and have one each. If they split the cost equally, how much is each hoodie?   _____

5   Maia cut a cake into 24 equal slices. She and her friend ate $\frac{3}{8}$ of the cake. How many slices were left?   _____

6   Hone has several single socks in his drawer. He has 3 black ones, 2 rugby socks, 5 school socks and 1 orange one. What is the probability that, without looking, he pulls out a school sock?   _____

Practice 3

1 **a** Melissa went to the supermarket and bought these items:
- 1 kg of grapes, at $6.50 per kilogram
- 2 bottles of juice, at $1.55 each
- a loaf of bread, $3.30 each
- 1.5 kg of ham, at $20 per kilogram.

How much change should Melissa receive from a $50 note? _____

b Her watch tracked her trip to the shop and produced this graph:

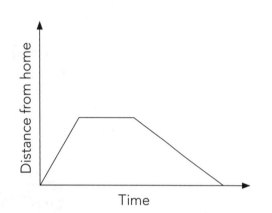

Which of these statements best fits the graph?

☐ She walked up the hill, did her shopping and ran down the hill to home.

☐ She walked northeast, did her shopping and then went home.

☐ She ran to the shop, did her shopping and then walked home.

☐ She walked to the shop, did her shopping and then ran home.

c She left home at 1439 hours and the trip to the shops and back took 35 minutes. At what time did she arrive home?

☐ 1474 hours ☐ 1504 hours ☐ 1514 hours ☐ 1574 hours

2 **a** Simon graphed the age and height of members of his family. Which graph represents the data in the table?

	Susie	Cam	Arnold
Age	5	10	15
Height (cm)	120	145	180

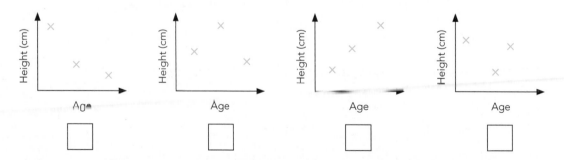

☐ ☐ ☐ ☐

b A year ago, Arnold was 1.67 m tall.
How much has he grown during the last year? _____ cm

3 Some tiles are missing from this pattern.

When complete, the pattern has two lines of symmetry. Which of these is the missing part of the pattern?

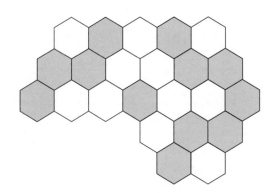

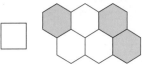

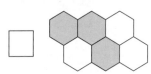

b Each hexagon tile has a perimeter of 18 cm. Which of these tiles has the same perimeter?

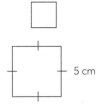
5 cm

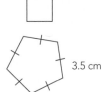
3.5 cm

6 cm

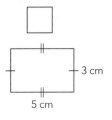
3 cm
5 cm

c Each tile costs $4.50. Calculate the cost of the tiles needed for the compete pattern.

$_____

4 The diagram shows the layout of the fairground, with the big tent near the centre.

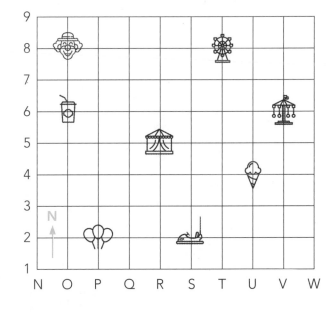

Key:
- Clown
- Ferris wheel
- Drinks stall
- Big tent
- Ice cream stall
- Bumper cars
- Carousel
- Balloons

a Write the coordinates for the ice cream stall. _____

b Eli is at the big tent and wants to see the clown. In which direction should he walk?

☐ NW   ☐ NE   ☐ SW   ☐ SE

Practice 4

1 One hundred Year 10 students wrote down the average number of hours (to the nearest hour) they spend exercising each day. The results are below.

Average daily exercise (hours)	0	1	2	3	4	5
Number of students	12	41	23	12	9	3

a Select the graph type that is most suitable for displaying this data.

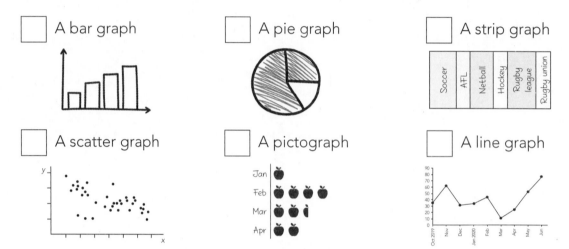

☐ A bar graph   ☐ A pie graph   ☐ A strip graph

☐ A scatter graph   ☐ A pictograph   ☐ A line graph

b Place a tick beside any true statements (there may be more than one):

☐ Over 90% of students did at least one hour's exercise per day.

☐ At least half the students did less than two hours' exercise per day.

☐ A quarter of the students did more than two hours' exercise per day.

☐ 10% of students did at least three hours' exercise per day.

2 a Rachelle is tiling her kitchen with this design:
What single transformation has she used to create this pattern?

☐ Reflection   ☐ Rotation   ☐ Enlargement   ☐ Translation

b Two tiles arranged like this are 15 cm long.
How many **tiles** will she need to complete a
1.8 m horizontal strip?

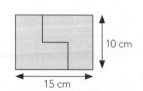

10 cm

15 cm

☐ 12   ☐ 24   ☐ 120   ☐ 240

c Calculate the perimeter of one tile.   Perimeter = _____ cm

3　**a**　The Smith family went on a holiday. At the end of their trip, the odometer (measures the total distance the car has travelled) in their car read: 
On their trip, they travelled 427 km.

0 6 4 8 3 5

What was the odometer reading at the start of their trip?

_____

b　The family left home at 8.41 in the morning and didn't reach their destination until 4.53 pm. How long were they travelling for?

☐ 7 hours and 12 minutes   ☐ 7 hours and 52 minutes   ☐ 8 hours and 12 minutes   ☐ 8 hours and 52 minutes

4　**a**　The school canteen has made up 5 L of hot chocolate drink. Each cup they sell contains 250 mL. If they sell seven cups of hot chocolate, how much is left?

☐ 3250 mL   ☐ 4750 mL   ☐ 0.325 L   ☐ 4.750 L

b　Hot chocolate drink is made by mixing hot chocolate powder with boiling water. This is the label on the back of a hot chocolate drink powder container:

Nutrition Facts

6 servings per container
Serving size　　**1 cup (230g)**

Amount per serving
Calories　　250

	% Daily Value*
Total Fat 12g	14%
Saturated Fat 2g	3%
Trans Fat 10g	11%
Cholesterol 8mg	3%
Sodium 210mg	9%
Total Carbohydrate 34g	12%
Dietary Fiber 7g	25%
Total Sugars 5g	
Includes 4g Added Sugars	8%
Protein 11g	

HOT Chocolate
since 1952   25¢
Let us warm you up!
OPEN 24 / 7

How many calories does the container hold?　　_____

c　How many grams of fat are there in two servings of chocolate milk?

☐ 28%   ☐ 28 g   ☐ 12 g   ☐ 24 g

Practice 5

1 The graph below shows Auckland's average number of sunshine hours per day for each month.

Average number of Sunshine hours per day

Hours (y-axis: 0, 2, 4, 6, 8)

Bars by month: January ≈7.5, February ≈7, March ≈6, April ≈5, May ≈4.5, June ≈3.5, July ≈3.7, August ≈4.6, September ≈5, October ≈5.6, November ≈6.5, December ≈6.6

Month (x-axis: January, February, March, April, May, June, July, August, September, October, November, December)

a In which months were the average sunshine hours below 4 hours per day?

☐ May, June, July and August   ☐ June and July

☐ July and August   ☐ Between April and September

b Next year, most days in Auckland will have less than 6 hours of sunshine.

☐ Agree   ☐ Disagree   ☐ Can't tell for sure

Explain your answer.

_____

_____

2 **a** Only 20% of ginger cats are female. If there are about 24 000 ginger cats in New Zealand, how many of these would you expect to be female? _____

b A cat sleeps for 16 hours a day.
For what fraction of the day is it awake? _____

c There are about 1.134 million cats in New Zealand. Write this number using numerals only. _____

3 **a** A jam recipe uses a ratio of 1½ cups of sugar for every cup of berries.
Tia has 5 cups of berries.

How many cups of sugar should she use to make jam? _____ cups

The table shows the mass of berries needed for the number of jars of jam.

Jars	Mass of berries (g)
1	150
2	300
4	600
8	1200

b Plot these points on the graph below and draw a line through them.
Extend your line to the edge of the graph in both directions.

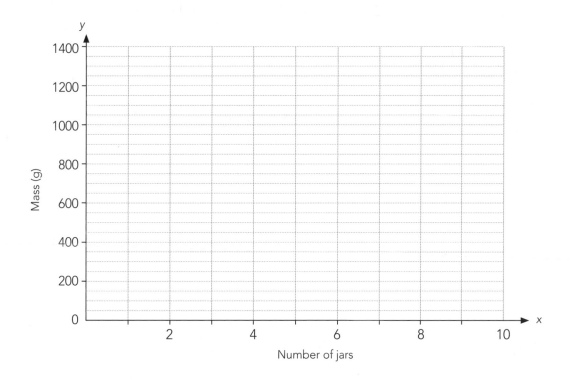

c Mark the point on the line where $x = 7$ and explain what this point shows.

_____

_____

d Berries cost $5.99 for a 125 g punnet.
How much would 1 kg of berries cost? $_____

e A 1.5 kg bag of sugar usually sells for $2.80. The supermarket
has reduced the price by 15%. Calculate the reduced price for
a 1.5 kg bag of sugar. $_____

Practice 6

1 The stars in Te Kakau (part of Orion) are shown below.

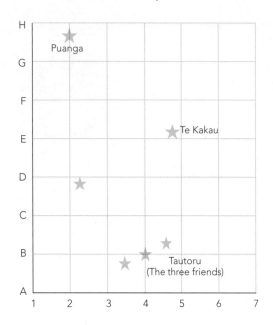

a Write the coordinates of the middle star of Tautoru (The three friends).

_____

b The ratio of the radius of the earth to the radius of the star Puanga is 1:8617.
Earth has a radius of 6371 km.

What is the radius of Puanga to the nearest million kilometres?

_____ km

2 A jug contains 1.6 L of milk. Two glasses of milk are poured. Each one contains 200 mL.

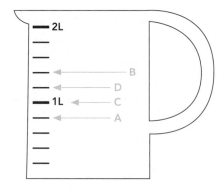

a Which arrow would show the amount of milk left in the jug?

☐ A   ☐ B   ☐ C   ☐ D

b These are the ingredients needed to make a batch of 20 scones:

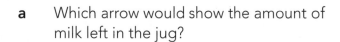

2 cups flour
¼ tsp salt
60 g butter
1 cup milk

Butter is sold in 500 g blocks.
What percentage of a block is needed to bake one batch of scones?

_____

c This scone recipe makes 20 scones.
How many cups of flour would be needed to make 50 scones?   _____ cups

3 **a** Gloria's bank statement is shown below.

Notice that bank statements sometimes have the oldest at the top.

Date	Description	Credits	Debits	Balance
13 Jun 22				$820.81
14 Jun 22	Supermarket		$54.75	
15 Jun 22	Pay	$598.40		$1364.46
15 Jun 22	Dairy		$12.49	$1351.97
16 Jun 22	Car repair		$253.99	$1097.98

What value is missing from the grey box? $_____

b $47.78 is deposited into her bank account on 17 June.
Assuming no other credits or debits have been made, what is the new balance of her account?

☐ $773.03   ☐ $1145.76   ☐ $1050.20   ☐ $868.59

4 Aroha plots the number of siblings (brothers and sisters) that each member of her class has.

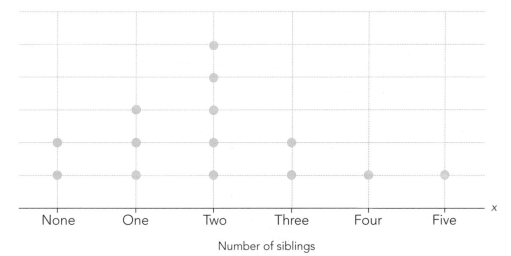

Number of siblings

a If a member of the class was picked at random, what is the probability that they have just one sibling?   _____

b What is the median number of siblings for these members of the class?   _____

c A new student joins the class. She has three siblings. Add this piece of data to the dot plot.

Practice 7

1 Huia is comparing potato prices at her local market.

    a Which of these is the cheapest per kilogram?

☐ $8.99 for 4 kg   ☐ $5.79 for 2.5 kg

☐ $2.99 for 1 kg   ☐ $4.80 for 2 kg

    b The logo on the potato bag has exactly two lines of symmetry. Which of these is the logo?

☐   ☐   ☐   ☐

    c The mass of potatoes grown one year in New Zealand was 525 031 tonnes.

        Write this number using words.

        _____

        _____

2 Simon needs to drive from Wakefield (a) to the Nelson Provincial Museum (b).

    a In which direction will he be driving?

☐ NW   ☐ NE   ☐ SW   ☐ SE

    b He needs to be at the museum at 11.10 am. If he allows
        5 minutes extra in case the traffic is bad, what is the latest
        time he can leave for the museum?   _____

3 The Wild Foods Festival was held for three days. This table shows how many people attended each day:

Day	Number of people attending
Friday	1286
Saturday	2791
Sunday	2385

a To the nearest hundred, what was the total number of people who attended the festival?

☐ 6400   ☐ 6460   ☐ 6000   ☐ 6500

b One of the popular things to eat at the festival are huhu grubs. The recommended daily protein intake for an adult is 165 g. Each huhu grub has a mass of 2.2 g on average.

How many whole grubs would you need to eat to meet this protein guideline?   _____

c Kara likes eating huhu grub lollies. Their heads can be red, orange, blue, yellow, green or purple. Her packet of lollies contains 5 red ones, 2 green, 4 purple and 1 blue.

If she closes her eyes and picks a lolly, what is the probability that she picks a red or a green one?   _____

4 In New Zealand, we use centimetres to measure length.
In the USA, feet are used.
2 feet is about the same distance as 61 cm (60.96 cm, technically).
If a table is 152 cm long, approximately how many feet is this?

☐ 2.5   ☐ 4.5   ☐ 5   ☐ 5.5

5 In New Zealand, wheelchair ramps can have a maximum rise of 1 metre for every 12 metres in base length.

Rise

Base length

What is the maximum rise of a ramp with a base length of 9 m?   _____ m

ISBN: 9780170474474

1   Josiah is considering buying a car.
The graph shows the relationship
between the ages and prices of
cars that he is considering.

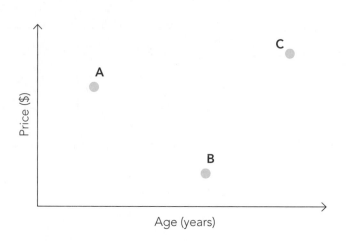

a   Place a tick beside any true statements (there may be more than one):

☐   The older the car, the higher the price.

☐   Car A is the youngest and the most expensive.

☐   Car B is cheaper and older than car C.

☐   Car C is the oldest and the most expensive.

b   He has $3000, but in order to buy his car he needs to borrow another $2500. He
will have to pay 14% interest on this amount each year. He repays the $2500 after
one year.

How much has the car cost him in total? $_____

c   Josiah's petrol gauge shows his tank is half full.
He spends $79 filling up his tank with unleaded 91.
These are the prices on the board:

To the nearest litre, how much petrol did he put into his tank?   _____ L

d   A car speedometer shows the speed at which
a car is travelling.

The speed limit is 100 km/hour.
What is the difference between the speed shown
on this speedometer and the speed limit?

☐ 11 km/hour   ☐ 13 km/hour   ☐ 21 km/hour   ☐ 23 km/hour

2 The population of New Zealand in 2018 was approximately:

	Population
North Island	3 594 552
South Island	1 104 537

a What percentage of New Zealanders lived in the North Island in 2018?

☐ 24%   ☐ 31%   ☐ 76%   ☐ 80%

b Round the South Island population to the nearest thousand.   _____

---

3 Emily wants to enlarge this photo.
The enlarged photo will be three times as wide and three times as high as the original.

What will be the area of the enlarged photo?

8 cm

6 cm

☐ 3 times the area of the original.

☐ 6 times the area of the original.

☐ 9 times the area of the original.

☐ 12 times the area of the original.

---

4 Maia is making cupcakes.
The first cupcake that she decorates is for the birthday boy, and it will have three chocolate buttons on top.
The rest will have two each.

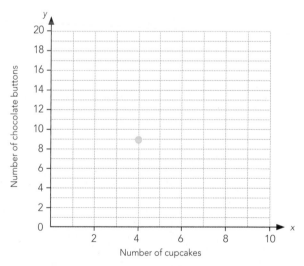

a The graph shows the relationship between the number of cupcakes and the number of chocolate buttons needed. Explain the meaning of the marked point.

_____

_____

b Add two more points to the graph and join the points to form a line.

Practice 9

1 Ben's bank statement is shown below.

Date	Description	Credits	Debits	Balance
10 May 22	Rent		$150.00	–$89.06
9 May 22	Supermarket		$67.29	$60.94
9 May 22	New TV		$1150.99	$128.23
8 May 22	Pay	$723.50		
7 May 22	Power		$260.00	$555.72

a What value is missing from the grey box?　　　　　$_____

b Ben is doing well financially.

☐ Agree　　　　☐ Disagree　　　　☐ Can't tell for sure

Explain your answer.

_____

_____

2 a Luca wants to paint the walls of his kitchen. What information will help him to calculate how much paint to buy?

☐ The volume of the kitchen.　　　　☐ The area of the walls.

☐ The perimeter of the room.　　　　☐ The capacity of the room.

b He needs to mix a shade requiring a green-to-white paint ratio of 3:1. How many litres of green and white paint would he need to make 20 litres of the new shade?

☐ 15 green and 5 white　　　　☐ 12 green and 8 white

☐ 14 green and 6 white　　　　☐ 10 green and 10 white

c He needs a ladder. The ladder he has chosen normally costs $230. However, because he found it in another shop at $229.99, he can get a reduction of 15% off the normal price.

How much will the ladder cost?　　　　　$_____

3 A family of two adults and two children are going skating.
A family pass would cost them $65.
Otherwise, an adult's ticket costs $24 and a child's ticket costs $13.

a How much do they save by buying a family pass? $_____

b The ice-skating rink logo is shown with possible
lines of symmetry.

Which are the lines of symmetry of this shape?

☐ B and D   ☐ B and C   ☐ A and C   ☐ A and D

4 Cam and his family are visiting the zoo. They are given this map so they can find their
way around.

Key

🐘 Elephants

🦁 Lions

🦌 Deer

🐢 Turtles

🐍 Snakes

🐦 Birds

🦭 Seals

a Write the coordinates for the turtles.   _____

b Cam is at the lions and wants to see the snakes. In which direction should he
walk?

☐ NW   ☐ NE   ☐ SW   ☐ SE

c Cam is standing at the elephant enclosure and is facing west.
If he turns 135° in an anticlockwise direction, which animals
would he be facing?   _____

Practice 10

1 The minimum and maximum temperatures each day in Queenstown were recorded for a week.

	Mon	Tues	Wed	Thu	Fri	Sat	Sun
Minimum temperature (°C)	–1	0	0	–3	–4	–3	–4
Maximum temperature (°C)	7	8	9	7	7	6	8

a Which day shows the greatest difference between the maximum and minimum temperatures? _____

b Queenstown is two hours ahead of Sydney's time. A plane leaves Sydney at 9.00 am. The flight takes three hours. What time is it in Queenstown when the flight lands?

☐ 2.00 pm   ☐ 12.00 pm   ☐ 10.00 am   ☐ 11.00 am

c The graph shows the relationship between New Zealand dollars and Australian dollars.

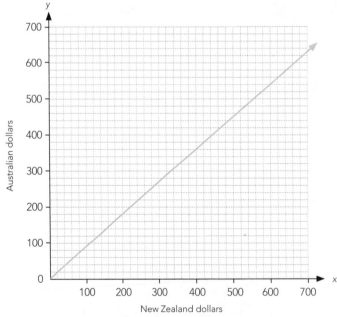

The ticket for the flight cost the passenger AU$540.00.

How many New Zealand dollars would be needed to buy this ticket?   NZ$ _____

d The maximum mass for a checked suitcase is 23 kg. Albert's empty suitcase has a mass of 4.3 kg.

Write the mass of the empty suitcase as a percentage of the maximum mass, and round your answer to 0 dp.   _____%

2 Manu has a lawn-mowing job. He can mow 9 m² per minute.

a How long will it take him to mow a lawn that is 558 m²?

_____ hour(s) _____ minutes

b A new customer has a lawn with the dimensions as shown.

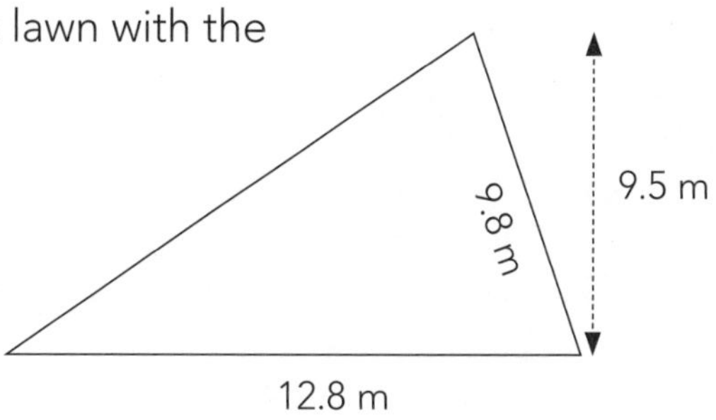

9.8 m

9.5 m

12.8 m

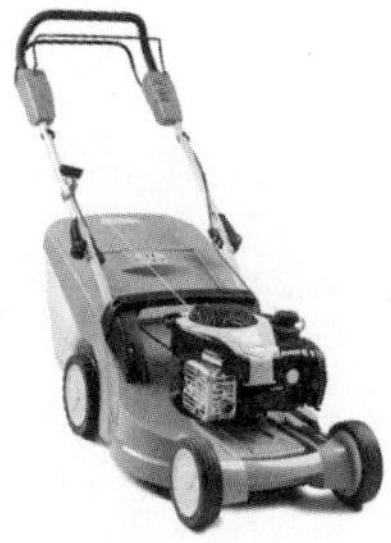

Which of these calculations would Manu use to work out the area of the lawn?

☐ (12.8 x 9.8) ÷ 2 ☐ (12.8 x 9.5) ÷ 2 ☐ (12.8 + 9.5) ÷ 2 ☐ (12.8 + 9.8) ÷ 2

3 **a** These are the results of a 200 m running race.

Lane number	1	2	3	4	5	6	7
Time (seconds)	26.23	27.12	26.20	24.56	26.25	26.61	26.12

Write the winning (fastest) time to the nearest second. _____ seconds

b The median of these times is:

☐ 26.25 seconds ☐ 26.61 seconds ☐ 26.23 seconds ☐ 26.20 seconds

c The athletes in lane 4, 5 and 6 were the fastest.

☐ Agree ☐ Disagree ☐ Can't tell for sure

Explain your answer.

_____

_____

d The symbol for the Colts Athletic Club is shown here.

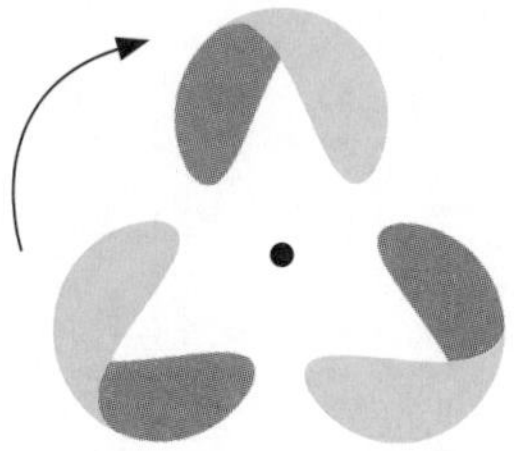

Through what angle has each 'C' been rotated? _____ °

Practice 11

1 A paddock has the dimensions below.

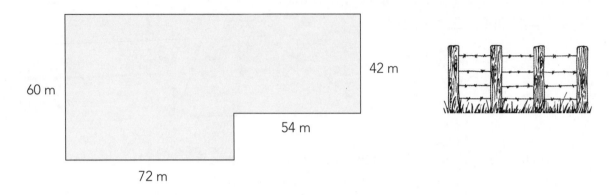

a Calculate the perimeter of the paddock. _____ m

b A fence is needed along the side which has a length of 72 m.
The distance between the centres of the posts must be 2.4 m.
Which of these calculations would be used to find the number of posts required?

☐ 72 ÷ 2.4 − 1   ☐ 72 ÷ 2.4 + 1   ☐ 72 x 2.4 − 1   ☐ 72 x 2.4 + 1

c It is recommended that fence posts have a third of their length underground. If the post is 1.8 m long, what length of post must be below the ground? _____ m

2 a There are 56 staff at Paradise High School. The ratio of females to males is 10 to 4. How many females would need to be replaced by males in order to make the ratio of females to males 1 to 1?

☐ 16   ☐ 6   ☐ 14   ☐ 12

b The staff were asked if they were left or right handed.

	Full time	Part time	Total
Left handed	9	3	12
Right handed	32	12	44
Total	41	15	56

What is the probability that a staff member is full time and left handed? _____

3 Lyn is at Central Station, and she must reach the Merrick Library no later than 8.30 am.

Bus stop	Time at bus stop (am)							
Central Station	6.40	6.58	7.10	7.35	7.47	7.54	8.05	8.16
Burke St	6.46	7.04	7.16	7.41	7.52	8.01	8.12	8.23
Greenvale	6.52	7.11	7.23	7.48	7.59	8.08	8.19	8.30
Reservoir	7.06	7.25	7.40	8.03	8.14	8.24	8.34	8.45
Merrick Library	7.17	7.37	7.50	8.14	8.25	8.36	8.45	8.56

a What is the latest bus she can take from Central Station? _____

b How long does the bus ride take? _____ minutes

c Every morning Lyn records the temperature at the bus stop.

Day	Temperature (°C)
Monday	11
Tuesday	9
Wednesday	10
Thursday	12
Friday	8

What was the mean temperature at the bus stop this week?

_____°C

d The bus company hired somebody to record whether their buses arrive at bus stops early, on time, or late. They recorded the times for 1000 arrivals.

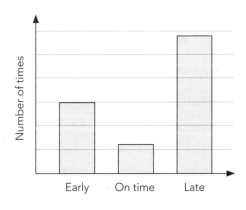

Estimate the number of times the bus was on time.

_____

e The bus was more often late than it was on time or early.

☐ Agree   ☐ Disagree   ☐ Can't tell for sure

Explain your answer.

_____

_____

Practice 12

1 The graph shows the price of a bag of carrots over a five-week period.

Price of one bag of carrots

a Place a tick beside any true statements (there may be more than one):

☐ Overall the price of a bag of carrots is increasing.

☐ The price of carrots was highest in week 5.

☐ The largest increase in price was between weeks 2 and 3.

☐ A bag of carrots was cheaper in week 3 than it was in week 5.

b During week 5, Abbie has to buy the carrots for the class camp. She needs 7 bags. How much change should she get from a $50 note? $_____

2 **a** Maya is organising a photo display for assembly. She has 32 photos and wants them to be displayed for 8 seconds each. How long will the presentation take?

☐ 4.16 minutes   ☐ 4 minutes 26 seconds   ☐ 4 minutes 16 seconds

b Maya's assembly could be on any day in Term 2. George has been unwell throughout Term 2 and stayed home on 16 out of the 48 days in Term 2.

What is the probability that he saw Maya's assembly slide show?   _____

c There is a total of 621 Year 9 and Year 10 students at Paradise high School. Each row in the assembly hall contains 32 seats, and there are 17 rows of seats.

Assuming all students are present, how many students will need to stand at the back?   _____

    ISBN: 9780170474474

3 Max and Marama are touring the South Island. They spent the night in Hokitika (bottom left) and they are going to drive along State Highway 6 to Kumara Junction.

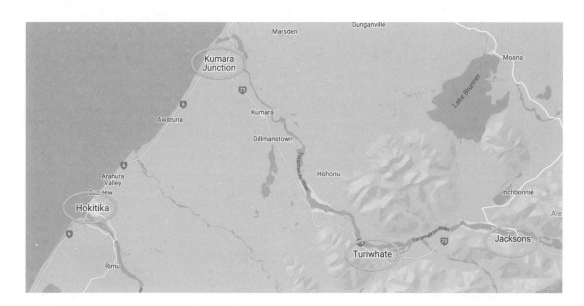

a It was 22 km from their motel in Hokitika to Kumara Junction. The trip took them 24 minutes. At what average speed (in kilometres per hour) were they travelling? _____ km/hour

b From Kumara Junction they will turn right onto State Highway 73 and drive to Turiwhate. In what general direction will they be driving? _____

c The total distance from their motel in Hokitika to Jacksons (far right of map) is 69 km.

This is their odometer reading when they left the motel:

What was their odometer reading when they reached Jacksons? _____

d Marama's car uses 8.2 litres for every 100 km they drive. Which calculation should she use to find out how much fuel she will use to travel 69 km?

☐ $8.2 \times \dfrac{69}{100}$ ☐ $100 \times \dfrac{8.2}{69}$ ☐ $69 \div \dfrac{8.2}{100}$ ☐ $8.2 \times \dfrac{100}{69}$

e Last January, Marama borrowed $1800 from her parents in order to help her buy the car. Each year she pays them 6% interest on what was owed, and she also repays $300 of the loan.

How much will she pay them next January? $_____

1 Cynthia has some tiles for the walls of her bathroom. Each edge of the tile is 30 cm long.

a She puts four tiles together to make this shape:

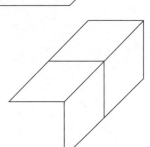

What is the perimeter of Cynthia's shape?

_____

b What transformation(s) could Cynthia use to get the blue tile to the position of the white one?

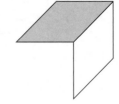

☐ Enlargement   ☐ Rotation   ☐ Reflection   ☐ Translation

2 Penny drives a truck that has a total height of 295 cm.

She drives it through a tunnel that has a ceiling 3.82 m above the road surface.

a How big is the gap between the top of the truck and the ceiling of the tunnel?

_____ cm

b She drove her truck 360 km in 4.5 hours. What was her average speed?

_____ km/hour

c In its annual report, her company boasted that in the last year they had doubled the number of trucks they own.

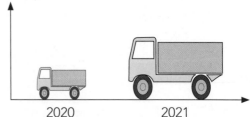

They used this graph to make their point.

Do you think this graph is a fair way to show this?

☐ Fair   ☐ Unfair

Explain your answer.

_____

_____

3 At the school camp, the staff know that each day they will need 2 L of milk for the staff plus 0.25 L for each student. This relationship is shown on the graph.

a Explain what point A tells you about the amount of milk needed.

_____

_____

_____

_____

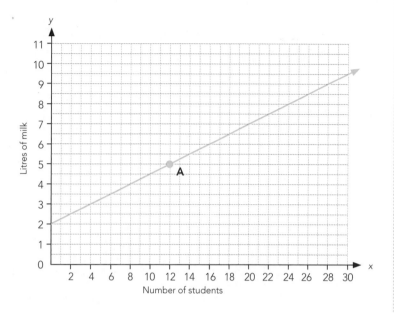

b How many litre bottles of milk will they need to buy for the first day if 26 students are coming to the camp? _____ L

c One teacher thinks that they should order 300 mL per student. Describe how this graph would change if they ordered 300 mL per student.

_____

4 This graph shows which sport a group of students want to play.

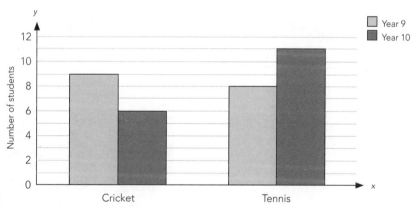

a How many students were in the group? _____

b Tennis is twice as popular as cricket with Year 10 students.

☐ Agree   ☐ Disagree   ☐ Can't tell for sure

Explain your answer.

_____

_____

Practice 14

1 Arlo was babysitting his little sister.
They built this object using 28 blocks.

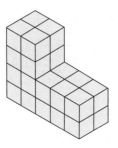

a Which of these has half the volume of Arlo's object?

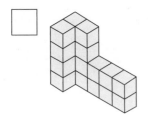   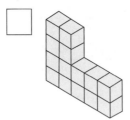   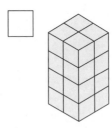   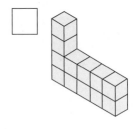

b After babysitting, Arlo went to the movies. The movie started at 7.40 pm and ran for 159 minutes. What time did it finish?

☐ 9.39 pm   ☐ 9.19 pm   ☐ 10.19 pm   ☐ 10.39 pm

2 a Two friends bought a pizza, which is cut into 12 slices. Luke ate one third of the pizza and Harvey ate a quarter. How many slices of pizza did they eat altogether? _____

b These are the dimensions of the pizza box (not to scale):

50 mm

36 cm   36 cm

The volume of this box in cm³ would be calculated using:

☐ 36 x 36 x 50   ☐ 36 x 36 + 5

☐ 36 x 36 x 5   ☐ 2 x (36 + 36 + 5)

c A 35 cm pizza normally costs $18.50. This week, all pizzas have been reduced in price by 12%. Calculate the reduced price of the pizza. $_____

ISBN: 9780170474474

3 A cake store receives orders in three different ways.

a

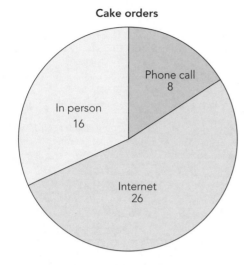

Cake orders

What percentage of the orders are made in person?

_____%

b A cake worth $28 is ordered. A delivery cost of 20% of the cake's value is added to the price. What was the total cost of the delivered cake?

$_____

c This is the design on the top of the cake:

Each koru is a clockwise rotation of the previous one.

By what angle does each koru need to be turned?

_____°

4 a How long is this piece of pounamu?

_____

b Bowenite is the most ancient type of pounamu. It is about half a billion years old.

Write half a billion using numerals. _____

1 These are the ages of the students in Daisy's gymnastics class.

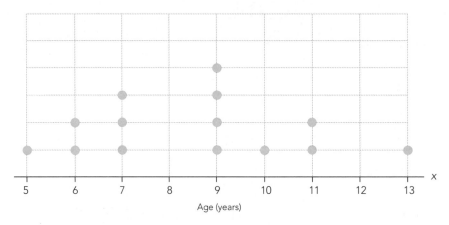

Age (years)

a What is the median age of the students in her class? _____ years

b Their evening recital starts at 4.30 pm and runs for 83 minutes.

At what time does it finish? _____

2 a Maakai is making curry. A supermarket sells four sizes of coconut milk. Which is the cheapest per millilitre?

$2.65

Coconut milk

400 mL

$3.10

Coconut milk

270 mL

$5.95

Coconut milk

1000 mL

$1.00

Coconut milk

165 mL

☐ A   ☐ B   ☐ C   ☐ D

b Maakai has 5 cups of rice. He uses 1½ cups for one meal and ¾ of a cup for another meal. How many cups of rice does he have left?

☐ 3¾   ☐ 2¼   ☐ 2¾   ☐ 2

c His rice recipe says he should use 1½ cups of water for every cup of rice. If he uses ¾ cup of rice, how much water should he use? _____ cups

3 The stars in Māhutonga (the Southern Cross) and Whetū Matarau (the Pointers) are shown below.

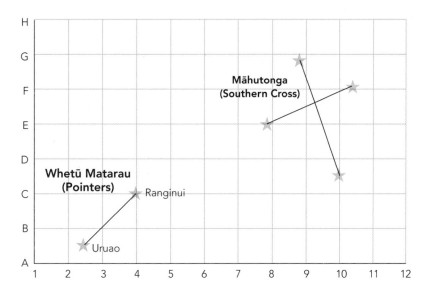

a Write the coordinates of Ranginui. _____

b Ranginui is about 14.1 million years old.
Write this number using numerals. _____

c One astronomical unit = the distance from the earth to the sun = 149 597 870 km.

Write the numeral that represents the number of hundred
thousands. _____

4 **a** Which of these nets would **not** fold to form a cube? _____

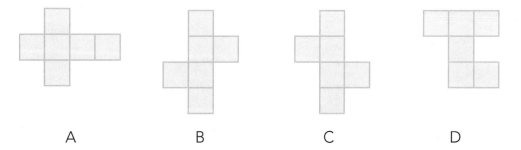

| A | B | C | D |

b All of these nets can fold to form cubes.

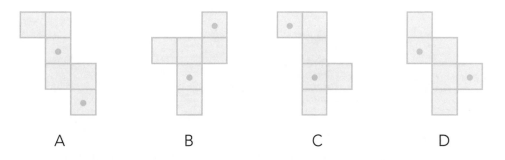

| A | B | C | D |

Which cube would **not** have dots on opposite faces? _____

Practice 16

1 **a** Anita is going on holiday. Her luggage has a mass of 36 kg. The limit for luggage is 23 kg, and every kilogram over this limit costs the flyer $15. Which calculation would be needed to find out how much Anita will have to pay for her excess luggage?

☐ (36 – 15) x 23   ☐ (36 – 23) x 15

☐ 36 x 15 – 23   ☐ 36 ÷ 15 – 23

b There were 258 passengers on her flight; $\frac{5}{6}$ of these were adults and the rest were children.

How many children were on the flight? _____

c The flight left at 10.43 am and landed at 12.36 pm. How long was the flight?

☐ 2 hours and 33 minutes   ☐ 1 hour and 93 minutes   ☐ 2 hours and 53 minutes   ☐ 1 hour and 53 minutes

2 **a** Maltilda designed a shape for a gift box. The box is placed on a grid.

Which of these shows the outline of the box on the grid?

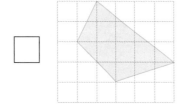 ☐   ☐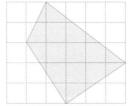

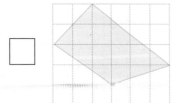 ☐   ☐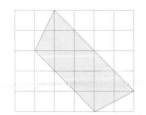

b The area of the blue face is 245 cm², and the height of the box is 8 cm.

Calculate the volume of the box. _____ cm³

3 a A survey was done of students' preferred breakfast food.

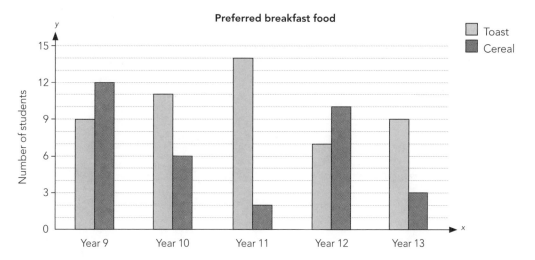

The majority of students prefer toast to cereal.

☐ Agree   ☐ Disagree   ☐ Can't tell for sure

Explain your answer.

_____

_____

b The recommended serving size for cereal is 33 g. How many
entire servings are there in a 1.5 kg family pack?   _____

c Lucinda is making muesli and needs 200 g of cashew nuts.
Roasted cashew nuts cost $24.90 per kilogram.
How much will 200 g of cashew nuts cost Lucinda?   _____

d She needs ¾ cup of oil for this recipe. She empties the oil bottle into a
measuring cup.

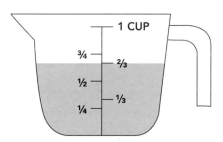

What fraction of a cup of oil does she
need to make this up to ¾ cup?

_____

e Her muesli needs to cook for 40 minutes. When Lucinda first put it in the oven,
this is what the digital clock looked like:

16:39

What time will it be when the muesli is cooked?

☐ 5.19 am   ☐ 5.19 pm   ☐ 17.19 am   ☐ 16:79

Practice 17

1   These are the plans for Samson's apartment. The lengths are in millimetres.

a   What is the internal area of his house?

☐ 4380 cm²   ☐ 43.8 cm²   ☐ 4380 m²   ☐ 43.8 m²

b   Estimate the width of the bed in metres.   _____ m

c   He bought a bulk box of soap. It contained 40 bars altogether: 12 of them were green, 18 of them were pink and the rest were white. If he picks one at random, what is the probability that it is green or white?   _____

d   Single bars of soap cost $2.25 each. His bulk box of 40 bars cost him $69.95.
How much did he save by buying 40 bars of soap in bulk? $_____

e   He has used 17 of the bars of soap. What percentage of his bulk box is left?   _____%

2   Carol's parents bought her a new phone but she has to pay them back.
The graph shows the rate at which she has to pay them.

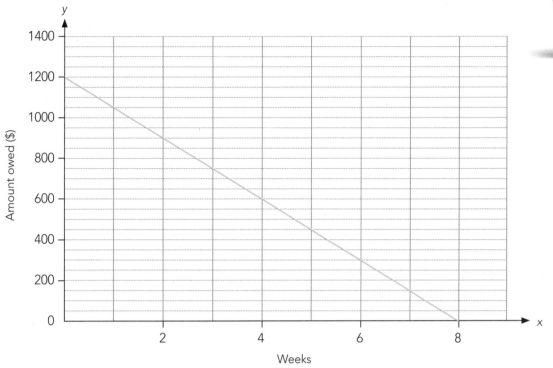

Weeks

a   How much does she pay back each week?   _____

b   How much does she still owe her parents after five weeks?   _____

c   Carol made a phone call that lasted 47 minutes.
The cost was $0.94 per minute.
She estimated that the phone call will cost her about $45.
This is a reasonable estimate of the cost of the phone call.

☐ Agree   ☐ Disagree   ☐ Can't tell for sure

Explain your answer.

_____

_____

3   **a**   In Omarama, the temperature was –4°C at 6 am. By 8 am
it had dropped another 2°C.
By midday it was 13 degrees warmer than it was at 8 am.
What was the temperature in Omarama at midday?   _____°C

b   The sun rose at 7.58 am and set at 5.07 pm. How
long was there between sunrise and sunset?   _____ hours _____ minutes

Practice 18

1   Tui and Rita are driving from their home in Martinborough to Featherston to visit their auntie for lunch. They will be travelling on State Highway 53.

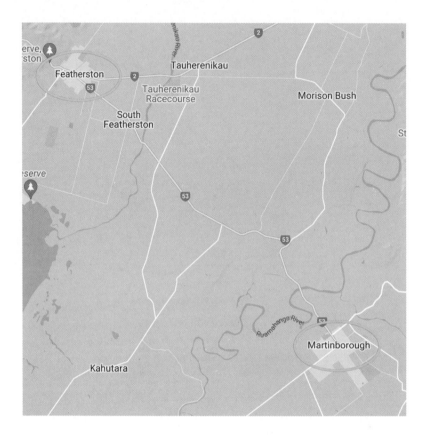

a   In what general direction will they be driving?   _____

b   The distance to their auntie's house is 17.5 km. If they average 70 kilometres per hour, how long will it take them to drive there?   _____ minutes

c   This is their odometer reading when you left home: `098095`

If they visited their auntie and returned home by the same route, what was the odometer reading when they returned?

_____

d   Rita's car uses 8.2 litres of fuel to drive 100 km, and they will drive 17.5 km each way.
How many much fuel will she use?   _____ L

e   Which distance is the closest to the maximum distance Rita can travel on one full tank (38 L) of petrol?

☐ 450 km   ☐ 500 km   ☐ 460 km   ☐ 380 km

2 Andre's English teacher got the class to vote on what novel to read in class.

Novel	Votes
The Hunger Games	12
The Boy in the Striped Pyjamas	5
Warrior Kids	8

a Which type of graph would be appropriate to show this data?

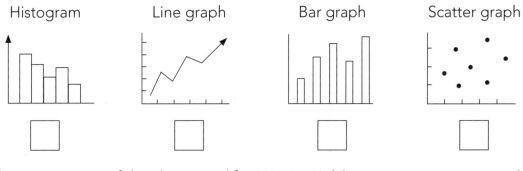

Histogram   Line graph   Bar graph   Scatter graph

☐   ☐   ☐   ☐

b What percentage of the class voted for *Warrior Kids*? _____%

c Andre has read 55 pages of his book. If he reads another 34 pages, he will have read half the book. How many pages are there in the book? _____

3 **a** Rupert is paid $35 an hour, but gets time and a half on Monday because it's a public holiday.
If he works for 6.5 hours on Sunday and 5.5 hours on Monday, which of these calculations would show how much he earns?

☐ 6.5 x (35 x 1.5) + 5.5 x 35   ☐ 6.5 x 35 + 6.5 + (35 x 1.5)

☐ 6.5 x 35 + 5.5 x (35 x 1.5)   ☐ 6.5 + 35 x 5.5 + (35 x 1.5)

b A customer comes into Rupert's work and wants to buy a pair of shoes.

Brand	Normal price	Sale price
Sprinters	$150	$120
Runners	$220	$165
Flyers	$180	$162
Joggers	$200	$170

She wants to buy the shoes which have the highest percentage discount. Which brand should she buy? _____

Practice 19

1   **a**   Students are competing in a triathlon that requires them to swim, bike and run.

Event	Distance
Swim	300 m
Bike	10 km
Run	3.4 km

What is the total length of the race?

_____ km

b   The two fastest male under-16 competitors recorded these times:

	Time (minutes:seconds)
First place	36:49
Second place	37:21

What is the difference in finishing times between these two competitors?

_____ seconds

2   The kapokapowai (New Zealand bush giant dragonfly) is shown below.

a

Estimate the wingspan (total width) of this dragonfly.

_____ mm

Scale: └_____┘
100 mm

b   Kapokapowai can fly at speeds of 56 km/hour. How long would it take one to cover a distance of 7 km?

_____ minutes _____ seconds

c   Kapokapowai spend the first $\frac{12}{13}$ of their lives as nymphs living in damp ground. Then the nymphs turn into dragonflies which live for 6 months. How long do they spend as nymphs?

_____ years _____ months

   ISBN: 9780170474474

3 This object was made using identical cubes.

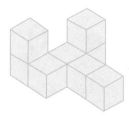

a Which diagram below shows the top view of this solid?

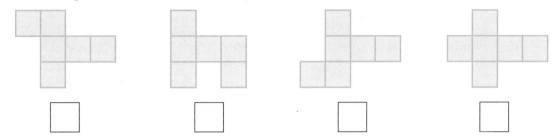

☐   ☐   ☐   ☐

b Each cube in this shape has an edge of 2 cm. Calculate the volume of the shape.

_____ cm³

4 The times taken for 16 children to finish a jigsaw puzzle were graphed.

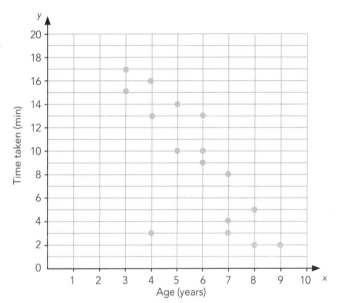

a Older children did the jigsaw faster than younger children.

☐ Agree   ☐ Disagree   ☐ Can't tell for sure

Explain your answer.

_____

_____

b There was one unusual point. Complete the sentence.

This child was _____ years old and took _____ minutes to do the jigsaw.

c Another child did the jigsaw puzzle. She was eight years old and took nine minutes. Add this point to the graph.

Practice 20

1 These are the ingredients for pancake batter:

1 cup self-rising flour
1¾ cups milk
2 large eggs

a Draw a clear arrow on the jug to show where 1¾ cups of milk should come to.

b It takes four large eggs to fill a cup.
What is the total volume of batter made by this recipe? _____ cups

Calculate the cost of making the pancake batter.

c 1.5 kg = 12 cups flour and costs $3.00.   1 cup flour: $_____

d 1 L milk = 4 cups and costs $3.20.   1¾ cups milk: $_____

e 12 eggs cost $7.20.   2 eggs: $_____

f   Total: $_____

2 A student scored the following marks in five tests:

Test 1: 16
Test 2: 17
Test 3: 19
Test 4: 11
Test 5: 16

Place a tick beside any true statements (there may be more than one):

☐ The median of her marks was 17.   ☐ The mode of her marks was 16.

☐ Her average (mean) was 16.0.   ☐ The range of her marks was 8.

3 Bernie was trying to draw a net for a rectangular prism.

C
D
B
A

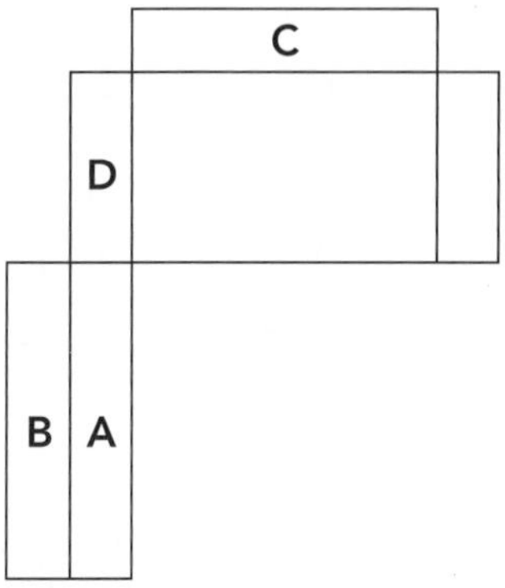

a He drew one of the faces incorrectly. Which one was it?

☐ A ☐ B ☐ C ☐ D

b The heights of all of these boxes are 0.8 m.
Which of them has a volume of 0.24 m^3?

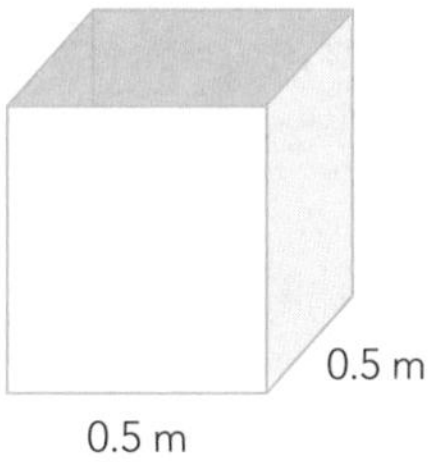

☐
0.5 m
0.5 m

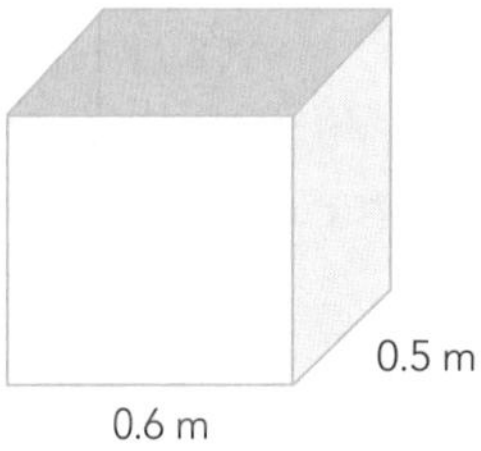

☐
0.5 m
0.6 m

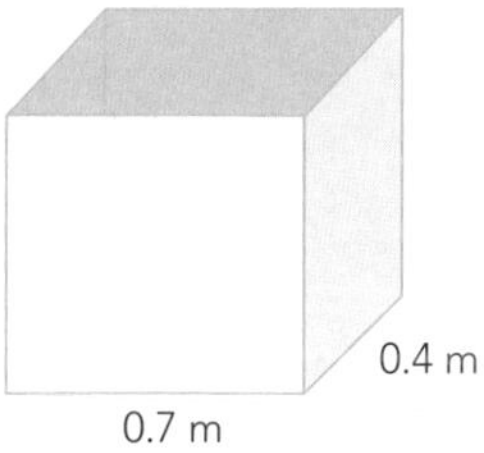

☐
0.4 m
0.7 m

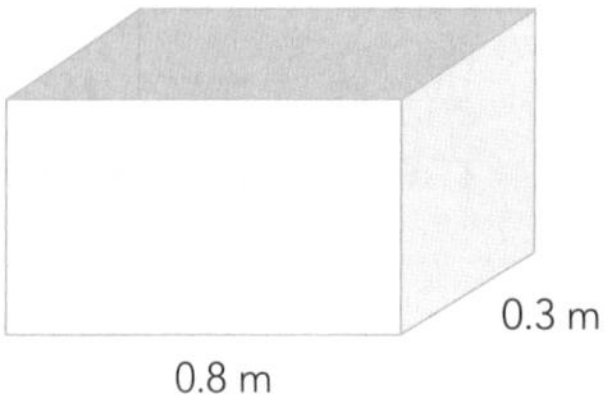

☐
0.3 m
0.8 m

c Matiu is going to use a box to market his new product. He would like the symbol on the box to have rotational symmetry greater than one. Which of these designs should he choose?

☐ ☐ ☐ ☐

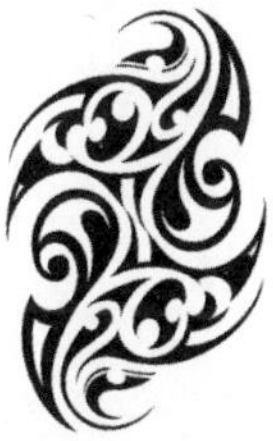

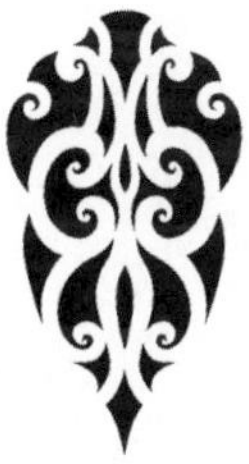

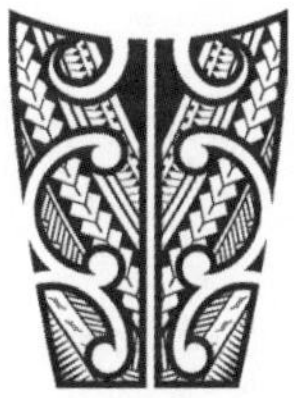

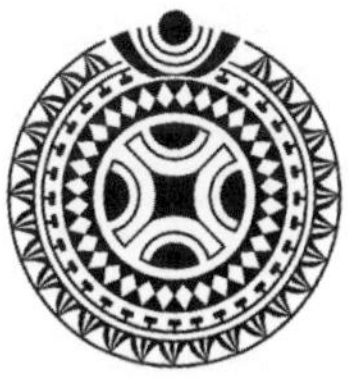

d His box will be a cube with edges of 0.2 m. What calculation is needed to find its surface area?

☐ 0.2^3 ☐ 6×0.2^3 ☐ 0.2^2 ☐ 6×0.2^2

ISBN: 9780170474474

Practice 21

1   Some students are participating in a step challenge for a month. They record the number of steps they walk each day.
This table shows the steps they have completed at the end of the 20th day.

Name	Steps
Rupert	214 048
Tia	243 503

a   To the nearest thousand, how many steps has Tia taken?

☐ 243 000   ☐ 240 000   ☐ 244 000   ☐ 214 000

b   Rupert's goal is to reach 320 000 steps by the end of the month.
With 11 days left, how many does he need to average each day
in order to make his goal?   _____

---

2   Each member of a class voted to decide who would be their council representative.

a   What is the probability that a class member voted for Aria?

_____

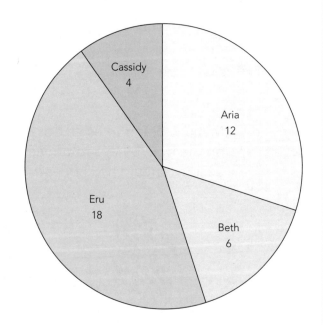

Cassidy 4
Aria 12
Eru 18
Beth 6

b   What percentage of the class members did not vote for Eru?

_____%

c   Which calculation should you use to calculate the angle of Beth's sector?

☐ 6 x 40 ÷ 360°   ☐ 6 ÷ 40 x 360°   ☐ 40 ÷ 6 x 360°   ☐ 4 x 6 ÷ 360°

3   This table shows the mass of some chickens and the time needed to roast them.

Mass	Time (min)
2 kg	100
1.5 kg	85
1 kg	70

a   Mark the points for each chicken on the graph and draw a line through the points. Continue the line to the edges of the graph in both directions.

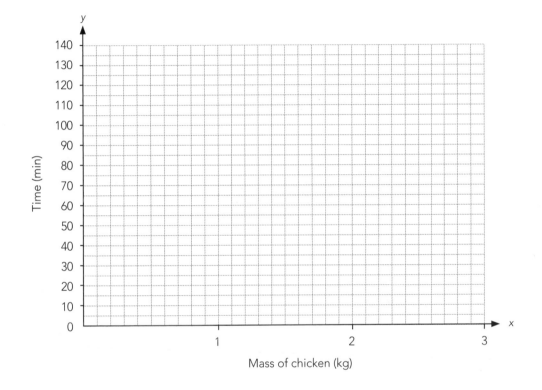

Mass of chicken (kg)

b   Mark the point on the graph which shows how long it will take to cook a 2.5 kg chicken.

c   Explain what this point shows about the length of time needed to cook a 2.5 kg chicken.

_____

_____

d   The supermarket sells the chicken for $8.60 plus 15% GST.
How much does a customer pay for the chicken? $_____

e   Some of these chickens are being bought for a school camp. One chicken serves 7 people.

If there are 33 non-vegetarians on the camp, how many
chickens will need to be bought?   _____

Practice 22

1 The star cluster Matariki is shown below.

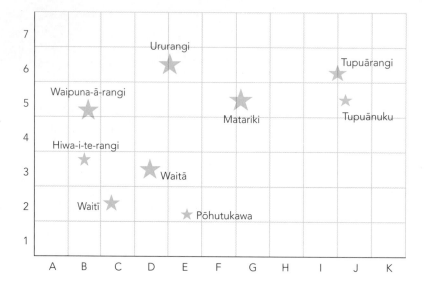

a Name the square containing Tupuānuku. _____

b The Matariki cluster is about 0.092 billion years old. Write this number using numerals only. _____

c There are about 1000 stars in the Matariki cluster, but only 14 of these are visible with the naked eye.

About what percentage of these stars are visible with the naked eye? _____%

2 Joseph participated in a three-day adventure race.

	Time (hours:minutes:seconds)
Day 1	6:24:31
Day 2	7:52:56
Day 3	5:09:18

a Which of these is Joseph's total time?

☐ 18:26:45   ☐ 19:25:05   ☐ 19:26:45   ☐ 18:85:105

b When he started on Day 2 the temperature was 8.2° but by the time he finished it was –3.7°. Complete the sentence:

The temperature had increased/decreased by _____ degrees.

3 **a**   It is recommended that one horse on its own needs 0.8 ha of area
for grazing, and that each additional horse needs another 0.4 ha.

How much area would eight horses need?   _____ ha

b   A rider has a mass of 64 kg and the mass of her horse is 0.7 tonnes. What is their
combined mass?

☐ 764 kg   ☐ 7640 kg   ☐ 70 064 kg   ☐ 7064 kg

c   The area needed for a horse in a paddock with others is 0.4 ha.
The ratio of the areas needed for horses to cows is 1:1.25.

What is the area needed for one cow?   _____ ha

4   Gilbert made this shape with some blocks.

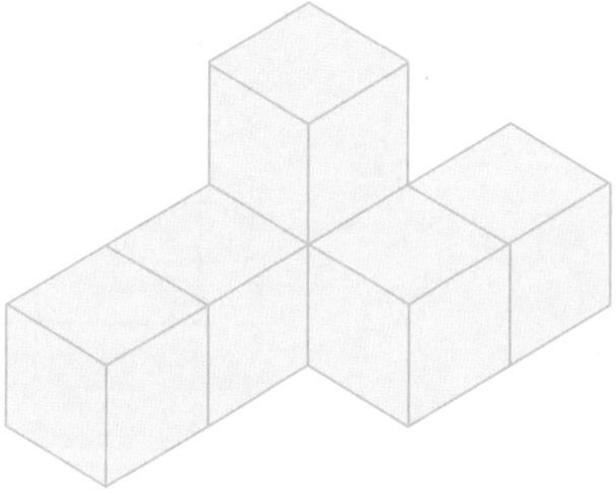

a   He sat his shape on some newspaper and spray-painted all the
faces he could see. How many faces would be painted?   _____

b   Which of these shapes is not the same as Gilbert's?

☐   ☐

☐   ☐

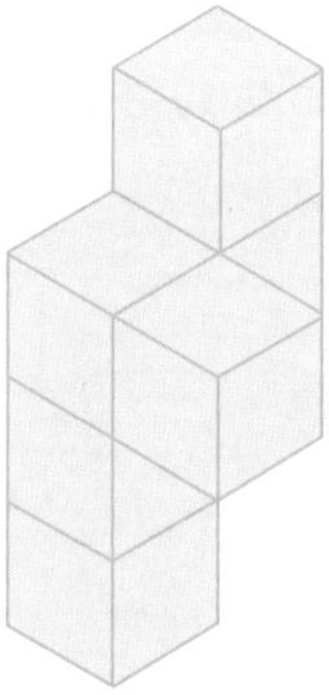

Practice 23

1 A family wants to ride a zip-line. The rules say that people must be at least 135 cm tall, and have a mass between 35 kg and 115 kg.

The graph shows the family's heights and masses.

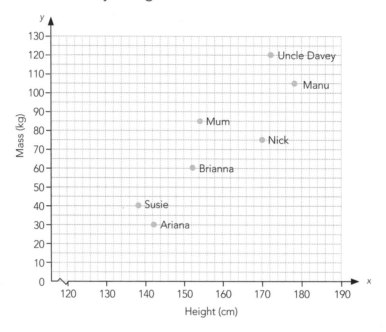

a Which of them will not be allowed to ride the zip-line? _____

b Place a tick beside any true statements (there may be more than one):

☐ Ariana is taller and heavier than Susie.

☐ Brianna is shorter than Mum, but heavier.

☐ Manu is taller than Uncle Davey, but lighter.

☐ Nick is taller and heavier than Brianna.

c These are the prices for the zip-line ride:

Family pass:	$337.00
Adult:	$129.00
Child:	$ 79.00

For 2 adults and 2 children, it is cheaper to buy a family pass than to buy separate adult and child tickets.

☐ Agree   ☐ Disagree   ☐ Can't tell for sure

Explain your answer.

_____

_____

2 At the school fair it costs $2 to have a spin of this spinner. You win the value the spinner lands on.

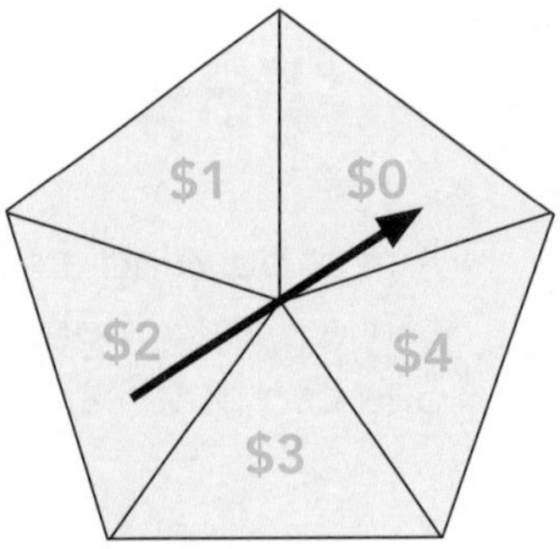

a Calculate the probability of getting nothing back.

_____

b Calculate the probability of getting at least $2 back.

_____

c The triangles forming the spinner are the same. Calculate the size of x, the angle at the centre of the spinner.

_____ °

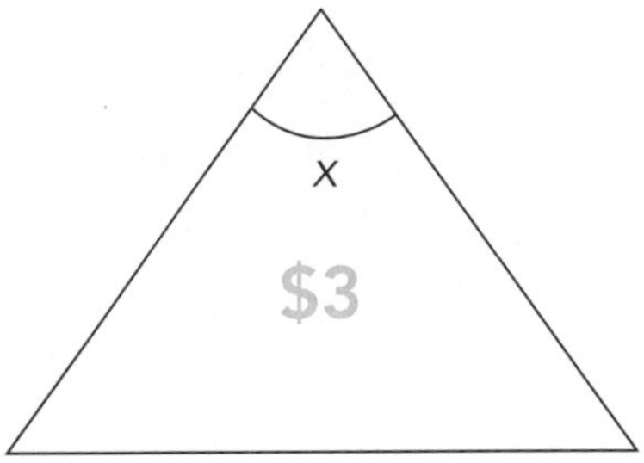

Craig wanted to check that the spinner was really fair. He spun it 200 times. These are his results.

$0	$1	$2	$3	$4
39	45	42	34	40

d The spinner is fair.

☐ Agree   ☐ Disagree   ☐ Can't tell for sure

Explain your answer.

_____

_____

e If he wanted to be more certain about whether or not it was fair, what should he do?

_____

_____

Practice 24

1 Kira is building a square garden bed that is
three planks high.
She can buy planks that are cut to the correct lengths.

a If planks cost $13.50 each, how much will Kira spend on planks
for this garden bed? $_____

b The inside of the bed is a square with sides 1.2 m long. The soil
will be 0.6 m deep.
Calculate the volume of soil needed to fill the bed. _____ m³

c To fill this bed along with another one, she needs 1.2 m³ of soil.
25 L bags of soil cost $8. There are 1000 L in a cubic metre.
Which calculation should she use to find the total cost of the soil?

$$\square \quad \frac{1.2 \times 1000}{25} \times 8 \qquad\qquad \square \quad \frac{25 \times 1000}{1.2} \times 8$$

$$\square \quad \frac{1.2 \times 1000}{25} \div 8 \qquad\qquad \square \quad \frac{25 \times 1000}{1.2} \div 8$$

d Kira keeps a thermometer in her garden.
When she read it at dawn, the temperature was –3°C. The
thermometer to the right shows the temperature at 10 am.

Describe the change in the temperature since dawn.

_____

e She recorded the temperatures at dawn every
morning for a week. These are her results in °C.

Mon	Tue	Wed	Thu	Fri	Sat	Sun
–4	–2	5	8	4	–3	–1

Which type of graph would be most appropriate
for showing this data?

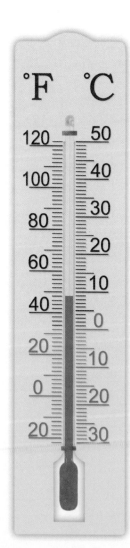

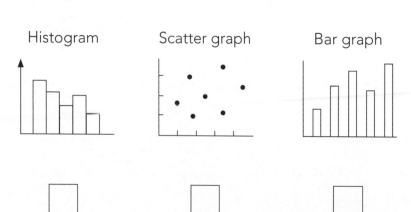

Histogram   Scatter graph   Bar graph

□   □   □

2 a A bath normally uses about 80 litres of water.
A shower uses 15 litres per minute.
What is the maximum length of time you could spend in the shower for your total water use to be no more than that for a bath?

_____ minutes _____ seconds

b The bath tap drips at a rate of one drip every three seconds.
10 drips = 1 mL.
Ari calculated that at that rate, 2.88 L of water would be wasted each day.
Show how he could have reached this conclusion.

_____

_____

_____

_____

_____

_____

3 Turkish Delight is a type of jelly-like sweet.
It consists of cubes with edges of 2.4 cm.
A box contains 25 cubes.

a Which calculation would you use to find the volume of Turkish Delight in the box?

☐ $2.4^3 \times 5^2$ ☐ $2.4^2 \times 5^2$ ☐ $2.4^2 \times 5^3$ ☐ $2.4^3 \times 5^3$

b The box is to have the maker's logo on the top. The logo must have an order of rotational symmetry of more than one. Which of these logos would be suitable (there may be more than one)?

☐   ☐

☐   ☐

1   Tom and his friends are staying at the campsite at Lake Rotoiti. They are going to walk the Loop Track in an anticlockwise direction.

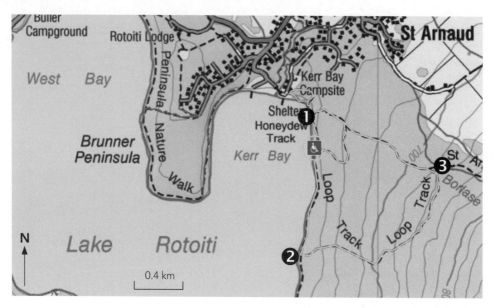

a   The first section of the track (❶ to ❷) goes along the lake shore. Use the scale on the map to estimate the length of this section.   _____ km

b   In what direction will they be tramping between point ❸ and point ❶?   _____

c   The Brunner Peninsula Nature Walk is 5.6 km long. It's a flat walk so they estimate that they will walk at 4 km per hour.

At that rate, how long will it take them to complete the Nature Walk? Give your answer in hour(s) and minutes.

_____ hour(s) and _____ minutes

2   a   Julia wants to ring her friend in Canberra, Australia, where they are 2 hours behind New Zealand. If Julia rings them at 1738 hours New Zealand time, what is the time in Canberra?

☐ 3.38 pm   ☐ 7.38 pm   ☐ 7.38 am   ☐ 3.38 am

b   The population of Canberra is 467 194.
Write this figure to the nearest thousand.   _____

c   Approximately 32% of the Canberra population were born overseas.
About how many of the 467 194 people were born overseas?   _____

3 Three friends went to the mall and spent $360 between them.

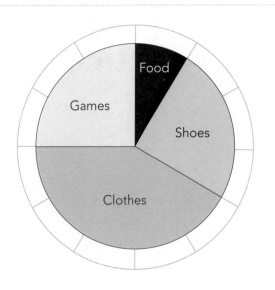

a How much did they spend on shoes?

$_____

b What percentage of their money did they spend on clothes and shoes?

_____%

4 Chris has designed this logo for her new company:

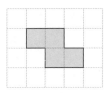

a Which diagram shows the logo enlarged by a scale factor of 2?

☐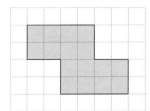

☐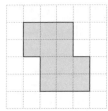

☐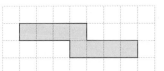

☐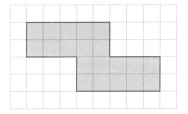

b Reflect her logo in the dashed blue line.

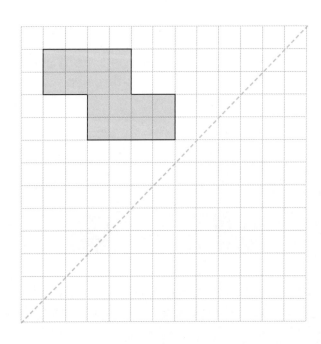

Practice 26

1 Tanya tracked the price of lemons at her local shop.

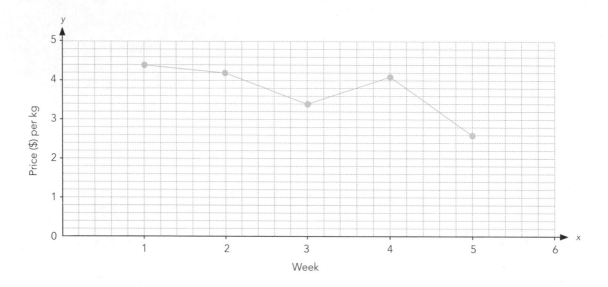

a Place a tick beside any true statements (there may be more than one):

☐ In week 1, the price of 1 kg of lemons was $4.10.

☐ The price of 1 kg of lemons dropped by 80 cents between weeks 3 and 5.

☐ The price of 1 kg of lemons was less than $4 in week 4.

☐ The largest drop in price was between weeks 2 and 3.

b The price of lemons in week 6 was $4.80 per kg. In week 7, the price increased to 12.5% more than it was in week 6. What was the price in week 7? $_____

2 **a** Hana picks a ball randomly from one of the bags below. Which one gives her a one-in-three chance of selecting a blue ball? _____

Bag A   Bag B   Bag C   Bag D

b If she pulls a white ball out of the bag, she wins a prize. Which bag should she pick so she is most likely to win a prize? _____

3 **a** Reggie is going to join a gym for 6 months.

Gym	Joining fee	Monthly charge
Push It	$45	$12
Vein	$0	$19
7 Days	$51	$10
Omega	$25	$15

Which of these gyms is the cheapest?

☐ Push It   ☐ Vein   ☐ 7 Days   ☐ Omega

b Reggie bought some new gear for the gym.

 $189    $58    $83

Show how he could estimate the total cost of these items.

_____

c Because he had joined the gym, Reggie was able to get 18% discount on the total for these three items. How much did they cost him? $_____

d Reggie is training for a running race which is 21 km long.
There will be six drink stations at evenly spaced intervals after the start of the race. How would you calculate where the last drink station should be?

☐ 21 − (21 ÷ 5)   ☐ 21 − (21 ÷ 6)   ☐ 21 − (21 ÷ 7)

e These are the times for the first 20 men crossing the finish line.

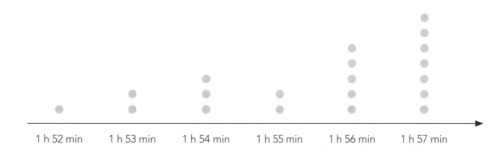

What was the median time for the first 20 runners? _____ hour(s) _____ minutes

f Reggie ran the race in 2 hours 31 minutes. What is the difference between Reggie's time and that of the winner?

_____ minutes

Practice 27

1 **a** Iriaka Rātana was the first female Māori MP. She was born on 25 February 1905 and died on 21 December 1981.

How old was Iriaka when she died? _____ years

b In 1949, there were only two parties in government: National and Labour.

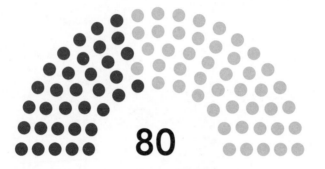

○ National

● Labour

80

What percentage of the 1949 government was made up of Labour members? _____%

c How else could this information be displayed?

Histogram	Line graph	Bar graph	Scatter graph
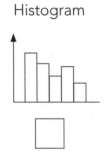		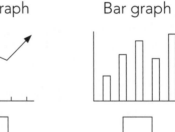	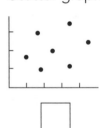
☐	☐	☐	☐

2 **a** A piece of timber is 1.75 m long. It needs to be cut into stakes that are 200 mm long.
How many stakes can be cut from the piece of timber?

☐ 17   ☐ 8   ☐ 9   ☐ 87

b Timothy has decided that the tops of the stakes need to be angled at 45°.

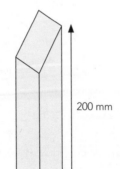

200 mm

Which will he be able to cut?

☐ More stakes than before.

☐ The same number of stakes as before.

☐ Fewer stakes than before.

3 **a** Poppy sold face masks for charity. Each face mask sold for $8. Plot this relationship on the graph below and join the points with a line.

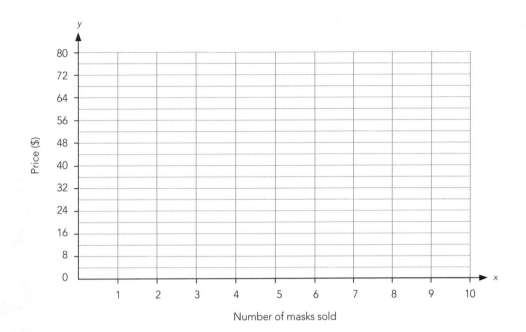

Price ($) (y-axis: 0, 8, 16, 24, 32, 40, 48, 56, 64, 72, 80)

Number of masks sold (x-axis: 1, 2, 3, 4, 5, 6, 7, 8, 9, 10)

b For each mask sold, $4.25 was donated to charity. If Poppy made $267.75 for charity, how many face masks did she sell? _____

4 **a** There is a 20% chance it will snow on Friday.
What is the chance that it will not snow on Friday?

☐ $\frac{1}{5}$　　　　☐ $\frac{4}{5}$　　　　☐ $\frac{1}{20}$　　　　☐ $\frac{20}{100}$

b A terrible snowstorm hit the USA in 2021. Another one had occured a century earlier.

In what year did the earlier snowstorm occur? _____

c Curtis measured the air temperature each hour.
• The first measurement was –0.4°C.
• The second was 2.1°C.
• For the third measurement, the air temperature had increased by triple the previous increase.

What was the third measurement? _____°C

Practice 28

1 a Fingernails grow on average 3.47 mm per month.

If your fingernails are very short, and you didn't cut your fingernails for a year, how long would they be in centimetres (provided they didn't break)?

_____ cm

b When painting nails, each normal-length nail uses around 0.0125 mL of nail polish.
Nail polish comes in 10 mL bottles.

If you did two coats per nail, which calculation would you use to work out how many nails you could paint with one 10 mL bottle?

☐ 10 ÷ (0.0125 x 10)   ☐ 10 x (0.0125 ÷ 10)

☐ 10 ÷ (0.0125 x 20)   ☐ 10 x (0.0125 ÷ 20)

c Toenails grow at a quarter of the rate that fingernails grow.

If Hamish didn't cut his toenails for six months, how many millimetres would they have grown?

_____ mm

2 The canteen sells milkshakes. The pie graph shows the popularity of the flavours.

a What is the probability that the next milkshake it sells is either caramel or vanilla?

_____

b Calculate the angle at the centre of the graph for the sector representing strawberry milkshakes. Round your answer appropriately.

_____ °

Milkshake flavours

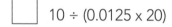

Vanilla 8%

Caramel 14%

Strawberry 32%

Chocolate 46%

3 **a** The diagram shows the net for a cube.
When the cube is made up,
which symbol would be opposite
the face with the **X**?

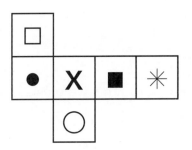

_____

b The net is rotated 90° in an anti-clockwise direction. Sketch the new view of this net, including the symbols on each face.

c Which of these diagrams shows another net which could be folded to form a cube?

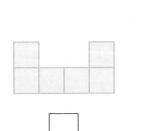   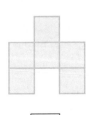      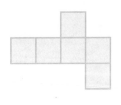

☐   ☐   ☐   ☐

4 **a** Greg wants to go skiing this year.

A day pass is $139.   A season pass costs $899.

If he goes skiing on seven days throughout the season,
would it have been cheaper to purchase a season pass?
Explain your answer.

_____

_____

b Hiring boots costs $15 per day and hiring skis costs $24 per day. Because he is a member of the club, he gets a 10% discount on hiring equipment.

How much will seven days' hire of boots and skis cost him? $_____

Practice 29

1 **a** Students were asked to record how long they slept on one particular night.

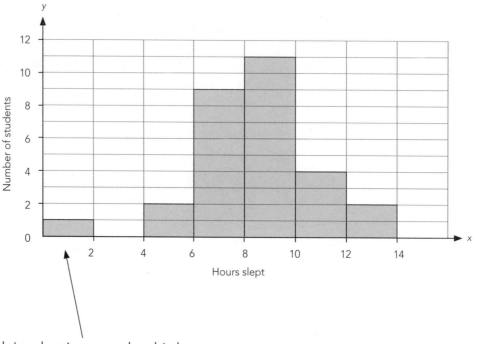

Explain what is meant by this bar. _____

_____

b How many students slept for at least eight hours? _____

c Hamish went to sleep at 2046 hours. He slept for 9 hours and 23 minutes.
What time did he wake up?

☐ 8.09 am   ☐ 7.09 am   ☐ 6.09 am   ☐ 6.69 am

2 **a** A school was providing morning tea for some visitors and needed to buy
sausage rolls.
These were the pack sizes and prices at the supermarket: 900 g for $8.00

800 g for $6.80

650 g for $5.50

Which size gave the best deal? _____

b There were 54 savouries put out. Some of the savouries were mince and some
were chicken, but both kinds looked the same.
Juliet took a savoury. She had a 4 in 9 chance of getting a chicken savoury.

How many mince savouries were there? _____

3 Travis made the pattern below.

a If he extends the pattern to 30 symbols, how many  symbols will he have used altogether? _____

b Which of these will be the 100th symbol in this pattern?

☐ ☺   ☐ ▲   ☐ ⬤   ☐ ✠

---

4 Ari packs mushrooms in the holidays. He is paid $20.50 per hour for a 40-hour week, plus time and a half for any extra hours that he works.

a Last week he worked 44 hours. How much was he paid? $_____

b The mushroms are packed into boxes containing between 1 and 4 kilograms.

What is the mass of the box on this scale?

_____ kg

c There are about five million species of fungi, of which about 14 000 are mushrooms.

About what percentage of fungi species are mushrooms?

Give your answer to 1 dp. _____%

1 Mum collected all the family's single socks and hung them on the line.

a If Ashton closes his eyes and chooses a sock at random, what is the probability that it has a cat on it? _____

b What is the probability that it has neither a star nor carrots on it? _____

2 a Which of these boxes has the largest volume? _____

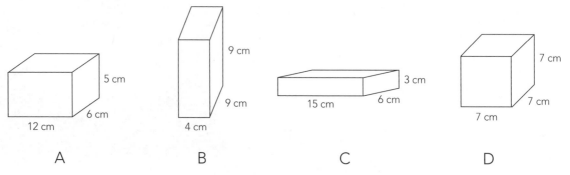

A   B   C   D

b The area of a square is 4 cm². What is the perimeter of the square? _____ cm

c Nathan is going to use box D for packaging a product. He will have the same symbol on all its faces. The symbol must look the same whichever way up the box is. Which should he choose?

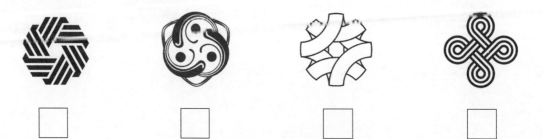

☐   ☐   ☐   ☐

3 The population in Stratford in the 2018 census was 5784. It was estimated to be growing by 6% each year.

    a Estimate the size of the population of Stratford exactly one year later in 2019. _____

    b The population of Hamilton in the 2018 census was one hundred and sixty thousand, nine hundred and eleven. Write this using numerals. _____

    c The bar graphs shows numbers of Hamilton people belonging to each age group.

Legend: ☐ Hamilton City ■ New Zealand

The population of Hamilton tends to be younger than the rest of New Zealand.

☐ Agree   ☐ Disagree   ☐ Can't tell for sure

Explain your answer.

_____

_____

4 **a** A killer whale (orca) will eat around 8% of its own body mass per day.

If a whale has a mass of 4 tonnes, what mass of food would it eat daily?

☐ 3.2 kg   ☐ 32 kg   ☐ 320 kg   ☐ 3200 kg

    b An orca can travel up to 56 kilometres per hour. At that speed, how long would it take to get the 182 km from Kaikōura to Christchurch?

_____ hours _____ minutes

Practice 31

1 Rudy went to the dog park and took note of the breeds he saw there.

Labrador	Pug	Poodle	Schnauzer	Border collie
12	4	2	3	7

a Select the graph types below that would **not** be suitable for displaying this data (there is more than one correct answer).

☐ Bar graph

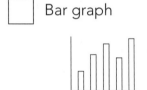

☐ Pie graph

☐ Strip graph

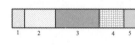

☐ Scatter graph

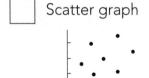

☐ Pictograph

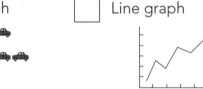

☐ Line graph

b Trudi noted the mass of her Labrador puppy as it grew.

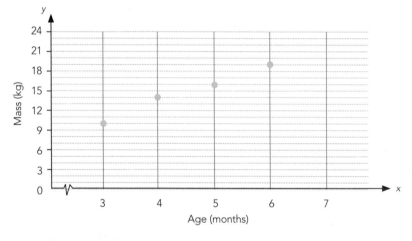

How much mass did her puppy gain between 3 and 4 months? _____ kg

c At 7 months old, her puppy's mass was 21 kg. Add this point to the graph.

d Trudi's dog grew to a mass of 30 kg.
Her dog needs 20 g Happy Dog food per kg of mass per day.
Happy Dog comes in 15 kg bags.
Trudi calculates that one bag will last her 25 days.
Show how she could have reached this conclusion.

_____

_____

_____

_____

_____

_____

ISBN: 9780170474474

2 **a** There are four teams in a round robin basketball tournament.
Each team must play every other team once. How many games
will there be in total? _____

☐ 4   ☐ 6   ☐ 8   ☐ 12

b The tournament finished at 5 pm.
Calculate the smaller angle between the two hands of this clock.

Angle = _____°

c The PTA sold cups of popcorn to the spectators.
Half a cup of corn makes 16 cups of popped corn. The PTA made
56 cups of popped corn. How many cups of corn did it use? _____ cups

d The cups cost 17.6 cents each.
The popped corn cost 4.4 cents per cup.
The PTA sold the 56 cups of popped corn for $2.00 each.
How much profit in total did the PTA make? $_____

e When Kara got home from the tournament, she felt bad. Her mum took her
temperature.

A normal temperature is 37°C. To the nearest tenth of a degree,
how many degrees higher than normal was Kara's temperature? _____°C

f Paracetamol is very good for bringing a temperature down. The normal dose
is 2 tablets, and no more than 8 tablets in total per day. How often can Kara take
2 tablets of paracetamol?

Every _____ hours

Practice 32

1 Below are the plans for Manu's new house. The lengths are in millimetres.

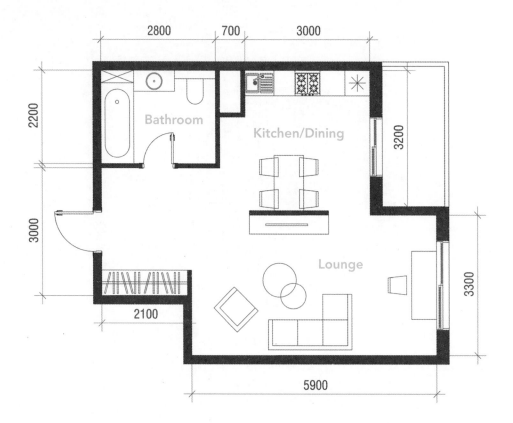

a Calculate the area of the bathroom. _____ m²

b Estimate the width of the large window in the lounge. _____

c Their front lawn will have the dimensions below.

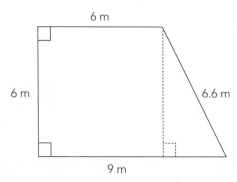

Which of these calculations would Manu use to work out the area of the lawn?

☐ $6^2 + \frac{1}{2}(3 + 6.6)$ ☐ $6^2 + \frac{1}{2}(3 + 6)$ ☐ $6^2 + \frac{1}{2}(3 \times 6.6)$ ☐ $6^2 + \frac{1}{2}(3 \times 6)$

d The back lawn has an area of 33 m².
800 g of lawn seed is enough to cover 24 m² of lawn.
How many kilograms of lawn seed will Manu need for the
back lawn? _____ kg

2 **a** This pie graph shows students' favourite toast spreads.
Place a tick beside any true statements (there may be more than one):

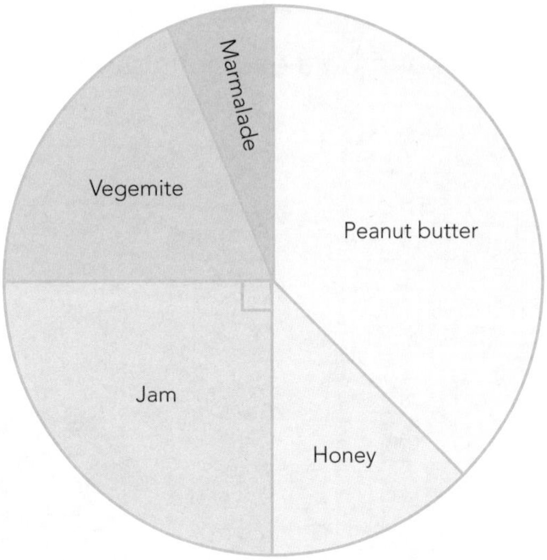

| | Honey was more popular than Vegemite. |

| | Less than a quarter liked Vegemite. |

| | Most of the students preferred peanut butter. |

| | The least popular spread was jam. |

b They surveyed 32 students. How many said they preferred jam? _____

c Ria estimates that the angle at the centre of the sector representing Vegemite is about 70°.

| | Agree | | | Disagree | | | Can't tell for sure |

Explain your answer.

_____

_____

3 **a** Pouākai (Haast's eagle) became extinct in about the year 1400.

To the nearest century, how many centuries ago was that? _____ centuries

b The ratio of the mass of a pouākai to that of a large seagull is about 17:1.
A large seagull has a mass of about 900 g.
Therefore the mass of a pouākai is about _____ kg.

c This ruler is drawn to scale. A reasonable measurement for the length of the beak of a pouākai is 12.7 cm. Draw a vertical line on the ruler to show this length.

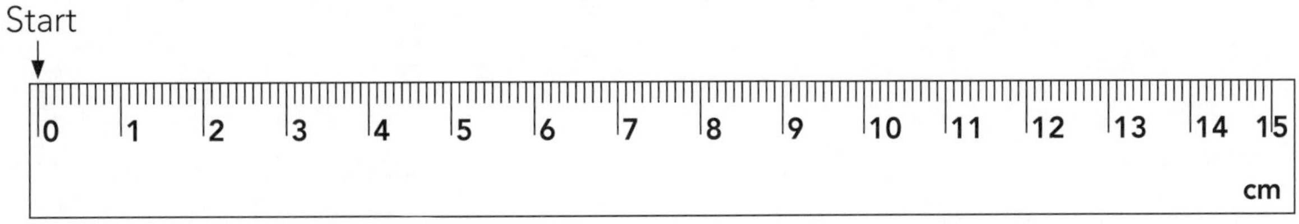

Start

1 A kiwi sanctuary has four species of kiwi. The numbers of sightings were recorded over three days.

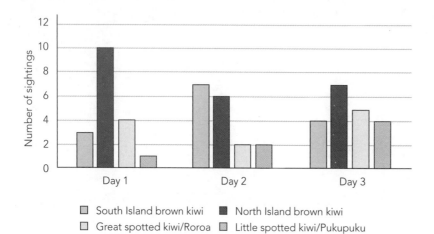

☐ South Island brown kiwi ■ North Island brown kiwi
☐ Great spotted kiwi/Roroa ■ Little spotted kiwi/Pukupuku

a On which day did you have the best chance of seeing all the species at the sanctuary?

_____

b Which species is a visitor most likely to see in the largest numbers?

_____

2 a Logan has bought himself a new bike. Its full price is $1199.00, but there is a sale which will reduce the price by 15%.

How much will he pay for the bike? $_____

b In order to buy his bike he has borrowed $600 from his parents. He has agreed to pay them 5% interest on this amount each year. After three years he pays back the entire sum and all the interest he owes them.

How much does he pay back? $_____

c He uses a pressure gauge to check that the pressures in his tyres are between 85 and 95 psi.
What is the reading on his pressure gauge?

_____ psi

3 a A bottle of hand sanitiser dispenses an average of 2.5 mL per pump.

A bottle contains 1.5 L. What is the maximum number of pumps you could expect to get from a bottle?

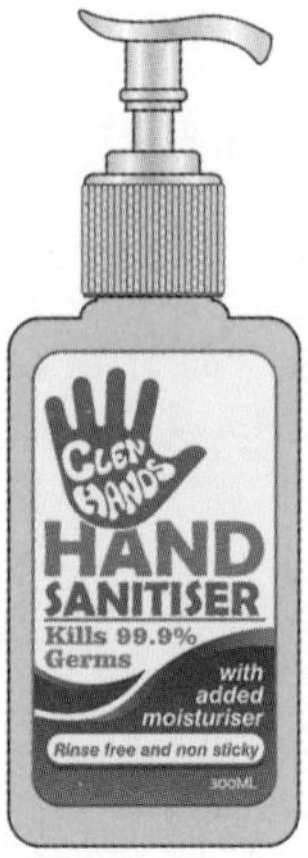

_____ pumps

b This hand sanitiser claims to kill 99.9% of 'germs'.
According to its claim, how many 'germs' does it not kill?

☐ 1 in 100   ☐ 1 in 1000   ☐ 1 in 10 000   ☐ 1 in 100 000

c Bacteria do not normally survive at temperatures above 60° Celsius.
In some countries all temperatures are given in degrees Fahrenheit.
To convert from Celsius to Fahrenheit you multiply by 1.8 and then add 32.
Keanu converted 60°C to Fahrenheit and his answer was 197.6°F.
Was he correct?

☐ Yes   ☐ No

If no, show how you would correctly calculate the temperature.

_____

_____

4 Sharee's payslip is shown below.

Income	Rate	Hours	Amount
Regular	27.50	79.5	$2186.25
Time and half		6	$247.50
		Total gross pay	$2433.75
Deductions		PAYE	$255.54
		Net pay	$2178.21

a 'Time and a half' means that the pay rate is 1.5 times the normal rate.
Show how the circled pay for time and a half was calculated.

_____

b PAYE stand for 'Pay as You Earn'. It is a tax which is subtracted from your total earnings.
Calculate the percentage (to 1 dp) of total gross pay that is PAYE.

_____%

Practice 34

1 **a** This table shows the mass and cost of kūmara.

Mass	Cost
500 g	$2.00
1 kg	$4.00
1.5 kg	$6.00

Mark the points on the graph below and draw a line through the points. Continue the line to the edge of the graph in each direction.

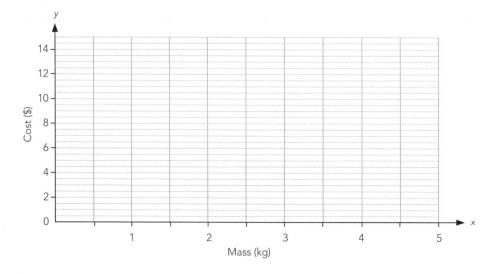

b Mark the point on the line where *x* = 2.5. Explain what this point shows about the cost of kūmara.

_____

c Tui is buying some kūmara plants for her garden. It is recommended that they are planted about 30 cm apart. Her garden bed is 2.6 m long. How many plants should she purchase?

_____

Explain how you got your answer. Include a diagram if you wish.

d Kūmara is $4.00 per kg.
What is the cost of a kūmara with mass 560 grams? $_____

e What mass is a kūmara that costs $1.30?

☐ 775 g   ☐ 130 g   ☐ 325 g   ☐ 520 g

2 a Steph does patchwork. She needs some more fabric and either of these would be suitable: 0.7 m of narrow fabric at \$5.60/m

or   0.6 m of wide fabric at \$7.50/m.

Which is cheaper and by how much? Show your calculations.

_____

_____

b Her patches have pictures of kiwis on them.

Which of these correctly shows the kiwi rotated by 270° in a clockwise direction.

A ☐   B ☐   C ☐   D ☐

c Steph stitches patches while she watches TV in the evenings.
These are the numbers of patches she has stitched during the last five evenings:

8   8   10   14   18

Which will change if 14 is removed? There may be more than one answer.

☐ Range   ☐ Mean   ☐ Median   ☐ Mode

3 Pekapeka (the New Zealand long-tailed bat) can fly at 60 km/hour.

a How long would it take a pekapeka to fly 200 m?

_____ seconds

b Adult pekapeka have a mass of about 9 g (about the same as a ballpoint pen). Newborn bats are about 3 g. If the ratio of adult to baby bat mass was the same as that for humans, what would be the mass of a baby born to a 75 kg woman?

_____ kg

1 This is part of a bus timetable.

	Marrow St	Daisy Drive	Supermarket	Forest Ave	Sports centre	Creek Rd
Time at bus stop	13:42	13:51	13:59	14:04	14:11	14:19
	13:57	14:06	14:11	14:16	14:23	14:31
	14:09	14:18	14:26	14:31	14:38	14:46
	14:26	14:35	14:43	14:48	14:55	15:03
	14:34	14:43	14:51	14:56	15:03	15:11

Katherine wants to get to the sports centre before 3 pm.

a What is the latest bus that she could from catch from Marrow St? _____

b How long does the bus ride take from Marrow St to the sports centre? _____ minutes

c The first bus on the timetable leaves Marrow St at 13:42. This can also be written as:

☐ 3.42 pm   ☐ 1.42 pm   ☐ 3.42 am   ☐ 1.42 am

2 Kiwifruit are exported in trays similar to the one in the picture below left.

The mass of each filled tray is about 3.7 kg.

The mass of an empty pallet weighs 48 kg. 256 trays are stacked on each pallet.

a Show how you could calculate the **approximate** mass of the pallet stacked with trays of kiwifruit.

_____

b A truck is licensed to carry no more than 36 tonnes of cargo. How many pallets, each with mass 1250 kg, can it carry?

_____ pallets

3 The graph shows the popularity of some baby names in New Zealand during 2019, 2020 and 2021.

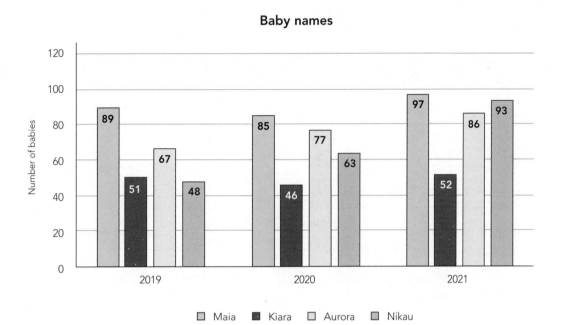

Baby names

Legend: ☐ Maia ◼ Kiara ☐ Aurora ◼ Nikau

a In which year was Nikau the least popular of these four names? _____

b In 2022 a baby girl is more likely to be called Maia than Aurora.

☐ Agree   ☐ Disagree   ☐ Can't tell for sure

Explain your answer.

_____

_____

c There were about 30 000 boys born in New Zealand in 2021.
What percentage of them were called Nikau? _____%

d A total of 58 659 babies were born in New Zealand in 2021.
Of these, 46,299 were born in the North Island. To the nearest
hundred, how many babies were born in the South Island
of New Zealand in 2021? _____

e The population of New Zealand when rounded to the nearest million is 5 million.
Place a tick beside the numbers that could be the actual population of New
Zealand (there may be more than one).

☐ 4 499 867   ☐ 5 501 347   ☐ 5 083 426   ☐ 4 506 459

Practice 36

1 The table shows the percentages of New Zealand adults in each activity level.

Indicator		2020	2021
Highly physically active	Did at least 5 hours of activity in the last week.	19.9%	20.7%
Physically active	Did between 2.5 and 5 hours of activity in the last week.	23.8%	24.3%
Insufficient physical activity	Did between 30 minutes and 2.5 hours of activity in the last week.	33.7%	33.2%
Little or no physical activity	Did less than 30 minutes of activity in the last week.	22.6%	21.8%

a In 2020, what is the probability that a randomly selected adult was either physically active or highly physically active? _____

b New Zealand adults were doing more physical activity in 2021 than in 2020.

 Agree   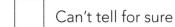 Disagree   Can't tell for sure

Explain your answer.

_____

_____

2 Eric has the choice of four spinners. They are all fair – they have equal chances of landing on any sector.

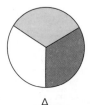   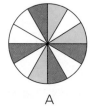   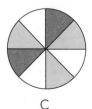   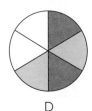

A   A   C   D

a Which spinner has one chance in four of landing on a grey sector? _____

b If a spinner has to land on white for a win, which spinner should he choose? _____

c Eric spun spinner D five times and every time it landed on white. His friend said his next spin must be grey or black.

 Agree   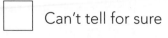 Disagree   Can't tell for sure

Explain your answer.

_____

_____

3 Annie and Tia both need a new kettle. They have both found the model they want at a normal price of $169.90 in two different shops.

Alf's Appliances is advertising a 'Buy one, get one half price' deal.

Best Bargains has reduced all prices by 30%.

a How much will they have to pay for each kettle if they buy them both from Alf's Appliances? $_____

b Calculate the price of a kettle if they buy from Best Bargains. $_____

c At Davy's Deals the same kettle costs $117.95. If Tia pays cash, the price will be rounded to the nearest 10 cents.

How much will she have to pay? $_____

d Tia is using her new kettle to brew coffee. She has discovered that the best temperature for brewing coffee is 92°C.

Draw a clear line on this thermometer to show where the needle should point for 92°C.

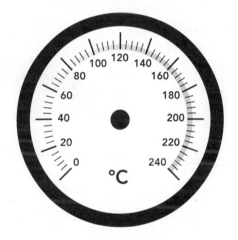

e Ground coffee comes in 200 g packs which cost $8.80 each.
One cup of coffee needs 35 g of ground coffee.
Tia has correctly calculated that each cup will cost $1.54.
Explain how she reached this conclusion.

_____

_____

_____

_____

_____

_____

_____

Practice 37

1 **a** An annual pass to the zoo costs $89.00. A day pass costs $36.50. How many times do you have to go during a year to make the annual pass worthwhile? _____

   b One of these zoo signs has one line of symmetry. Which is it?

☐ ☐ ☐ ☐

   c The zoo has tuatara. Tuatara relatives have existed for 0.22 billion years. Write this number using numerals only.

   _____

   d Use the scale to estimate the total length of this tuatara. _____

20 cm

2 Tim visited his local fruit and vegetable shop and bought carrots, lettuce, pumpkin and potatoes.

   a Lettuces in the supermarket were $4.50. What percentage (to 0 dp) of the supermarket price did he **save** by shopping at Matiu's Fruit and Veg?

   _____%

   b At the supermarket, 5 kg bags of potatoes were $17.95.
   How much cheaper were they per kilogram at Matiu's Fruit and Veg? Show your calculations.

   _____

   _____

Matiu's Fruit and Veg		
3 Toru St		
080 012 345		
CASH RECEIPT		
Carrots	1.25 @ 0.96	1.20
Pumpkin	1 @ 4.50	4.50
Lettuce	2 @ 2.99	5.98
Potatoes	3.2 @ 2.80	8.96
TOTAL		20.64
Cash		50.00
Change		29.36

3 Emily is sewing rhombus-shaped patches of fabric into this pattern.

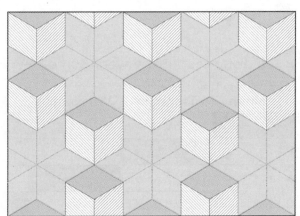

a Calculate the area of this patch.

Area = _____ cm²

b Write down the size of angle X.

Angle = _____°

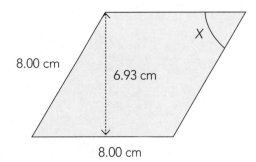

8.00 cm

6.93 cm

X

8.00 cm

4 Aroha recorded the temperature at 6.30 when she got up each morning. Here are her results for the days in July.

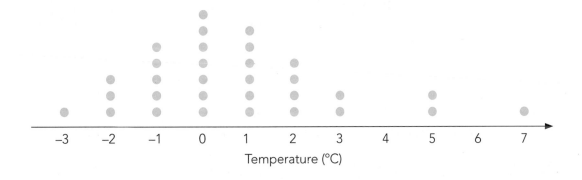

Temperature (°C)

a What is the median temperature at 6.30 each morning in July? _____°C

b There is a frost when the temperature drops below 0°C.
Her local newspaper headline says 'Coldest July in 40 years: frost on half of mornings'.

☐ Agree   ☐ Disagree   ☐ Can't tell for sure

Explain your answer.

_____

_____

Practice 38

1   **a**   This table shows the distances (km) between some New Zealand cities.

	Hamilton	Tauranga	Whakatāne	Rotorua	Taupō
Hamilton		107.4	191.3	108.1	152.6
Tauranga	107.4		87.6	62.8	149.8
Whakatāne	191.3	87.6		85.0	141.0
Rotorua	108.1	62.8	85.0		79.6
Taupō	152.6	149.8	141.0	79.6	

a   How far is it between Tauranga and Rotorua?   _____ km

b   Koa travelled from Hamilton to Tauranga and then on to Whakatāne. How far did she drive?   _____ km

c   From Whakatāne she drove 85 km to Rotorua. The trip took her 1 hour and 15 minutes.

Calculate her average speed.   _____ km/hour

d   Later in her holiday she drove from Taupō to Hamilton. In what general direction was she driving?   _____

e   She bought an ice block in Hamilton. Its price was $2.95. How much change will she get from a $10 note? $_____

2 Millie is making the base for ginger crunch. The ingredients are shown below.

150 g butter	1½ cups rolled oats
⅔ cup sugar	½ cup coconut
¾ cup self-raising flour	2 tsp ground ginger

a She is measuring the sugar.

Draw a clear arrow on the jug to show where ⅔ cup of sugar should come to.

b The recipe says that the mixture should be pressed into a baking tin that is 21 cm by 27 cm.

Milly uses her 24 cm square tin. This means that the mixture will be

☐ slightly thicker than the recipe says.

☐ the same thickness as the recipe says.

☐ slightly thinner than the recipe says.

c The base of ginger crunch needs to be cooked for 25 minutes. When she puts it in the oven, its 12-hour clock reads 12:47.

What will it read when the base is cooked? _____

d After she has iced the ginger crunch she cuts it into rectangular pieces which are each 3 cm wide and 8 cm long. How many pieces will she get? _____

e The butter cost $5.75 for the 500 g block. Calculate the cost of the 120 g of butter needed for this recipe. $_____

1   Beth has stuck these blocks together to make this shape.

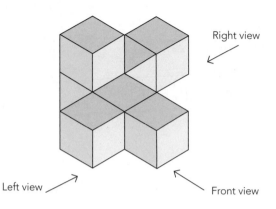

Right view

Left view   Front view

a   Which diagram below shows the right view of this shape?

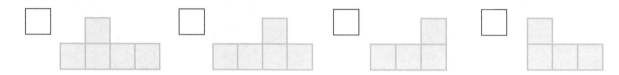

b   The completed shape was sat on some newspaper and sprayed in yellow paint.
How many faces of the blocks will be yellow?

☐ 21

☐ 22

☐ 23

☐ 24

c   Beth added more blocks to her shape to make a cuboid that was two blocks high, three blocks deep and four blocks long.

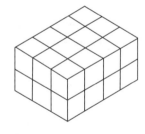

How many **more** blocks did Beth need?

_____ blocks

2   a   Grass grubs feed on the roots of plants and cause damage to farmland grasses and lawns during the months of January through to October (inclusive). For what percentage of the year do grass grub **not** damage pasture? Give your answer to 1 dp.

_____ %

b   A paddock which is 260 m long and 80 m wide has an average grass grub density of 200 per square metre. Approximately how many grass grubs does it contain?

_____

3 Toby is training for a triathlon. The strip graph shows how much time he spends on each type of training.

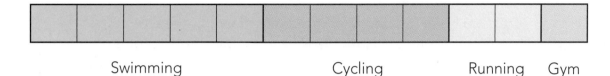

Swimming   Cycling   Running   Gym

a If he spends a total of 18 hours a week training, how long does he spend swimming?

_____ hours

b The average speed for the swimming phase of the triathlon was 100 m in 2 minutes.
Change this average speed into kilometres per hour, showing your reasoning.

_____

_____

_____

c A triathlon bike wheel rotates through 1800° every 10 m. How many full rotations is this?

_____ rotations

d The bike stage of a triathlon is 40 kilometres. Toby did it in 1 hour and 30 minutes. What was his average speed? Round your answer to the nearest km/hour.

_____ km/hour

e As part of his training, Toby bikes around a circuit from Darfield to Glentunnel to Hororata to Charing Cross and back to Darfield.

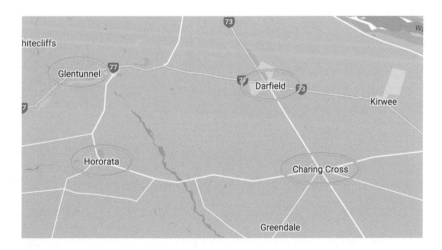

In what general direction is he riding between Charing Cross and Darfield?

_____

Practice 40

1 **a** Curtis and his friends went fishing during a long weekend.
The earliest Curtis can launch his boat is one and a half hours after low tide.
If low tide is at 13:54, what is the earliest time he can launch his boat?

☐ 14:54   ☐ 14:24   ☐ 15:54   ☐ 15:24

The graph shows what fish they caught each day.

(Bar graph titled "Number caught" on vertical axis (0–9), with Day 1, Day 2, Day 3 on horizontal axis.)

Legend: ☐ Snapper   ☐ Gurnard   ☐ Blue cod   ☐ Terakihi

b Which fish did they catch the most of? _____

c Curtis says that the best day of fishing was on the second day.

☐ Agree   ☐ Disagree   ☐ Can't tell for sure

Explain your answer.

_____

_____

d In New Zealand, approximately 2400 tonnes
of blue cod are harvested each year.
The average mass of a blue cod is 800 g.
Approximately how many blue cod are
harvested each year?

_____

e They caught nine blue cod during the three days. These were their lengths (cm):

33   34   35   35   38   39   40   42   46

Place a tick beside any true statements (there may be more than one).

☐ mean > median > mode   ☐ mean = median > mode

☐ mean > median = mode   ☐ mean = median = mode

              ISBN: 9780170474474

2 **a** The short sides of the triangles within this rectangle are 3 cm long.
Calculate the area of the rectangle.

_____ cm²

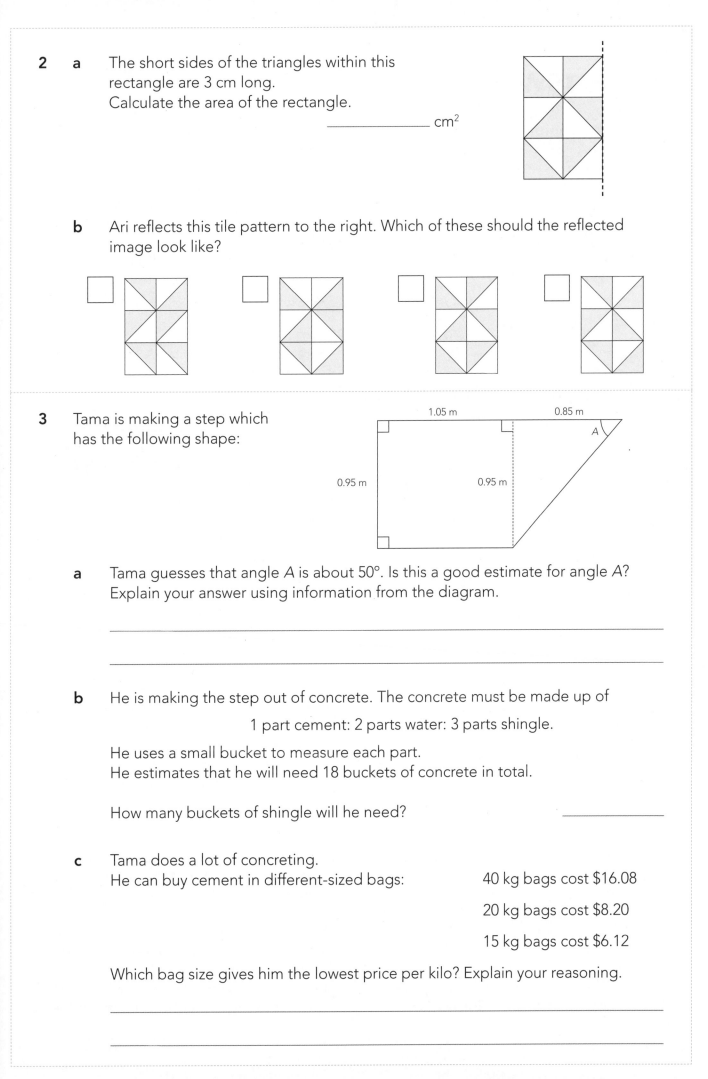

b Ari reflects this tile pattern to the right. Which of these should the reflected image look like?

☐   ☐   ☐   ☐

3 Tama is making a step which has the following shape:

1.05 m   0.85 m

A

0.95 m   0.95 m

a Tama guesses that angle A is about 50°. Is this a good estimate for angle A?
Explain your answer using information from the diagram.

_____

_____

b He is making the step out of concrete. The concrete must be made up of

1 part cement: 2 parts water: 3 parts shingle.

He uses a small bucket to measure each part.
He estimates that he will need 18 buckets of concrete in total.

How many buckets of shingle will he need?   _____

c Tama does a lot of concreting.
He can buy cement in different-sized bags:

40 kg bags cost $16.08

20 kg bags cost $8.20

15 kg bags cost $6.12

Which bag size gives him the lowest price per kilo? Explain your reasoning.

_____

_____

1 Formulate approaches to solving problems (pp. 6–7)

1 Deal 1: Four plants will cost 22.50 x 2 = $45.00
Deal 2: Four plants will cost 4 x 12.95 x 0.8 = $41.44
Deal 2 is cheapest for four plants by $3.56.

> 0.02 x 12 because there are 12 months in a year.

2 0.02 mm per month = 0.24 mm per year.

0.24 mm per year = 1000 mm in $\frac{1000}{0.24}$ years

> 1 m = 1000 mm

= 4167 years (1 dp)

> Divide 1000 by 0.24 to find the number of years taken to grow a metre.

3 She will need 225 x $\frac{72}{30}$ = 540 g of butter.

4 Multiply four and a half days by 24 (hours in a day)
= 108 hours
So it travelled 2160 km in 108 hours.
To convert this to km/hour, divide both sides by 108.
Speed = 20 km/hour.

5 The cost of hiring the digger is $90 plus $40 for each day.

2 Use mathematics and statistics (pp. 8–9)

1 a $x = 135°$ **b** $1750^2 – 50^2$

2 5000 steps

3 a 3.4, 3.5, 3.6, 3.7 or 3.8 m
  b 2 500 000
  c 4.6 kg
  d 45 minutes

4 a $3.80 **b** 30 cents

3 Explain the reasonableness of responses (pp. 10–15)

1 Agree
Because he is left with just $2.63 until he gets paid again.
Or:
Can't tell for sure
Because we don't know how often he gets paid or what other sources of income he may have.

2 Agree
However, this will overestimate the cost because the plant and bottle are both rounded up. As a result her estimate is about $6 more than the actual price.

Or:
Disagree
She would have been better to overestimate either the plant or the bottle and underestimate the other. If she had done this she would have got an estimate of $130 which is only about $1 more than the actual cost.

3 Disagree
Unfair because, although the height has doubled, so has the length. This means that the area of the 2022 bus is four times the area of the 2021 bus. It should have only doubled.

4 Can't tell for sure
Because we don't know how many people play rugby in each centre.

5 Disagree
Normal prices:
2 parents and 1 child = 2 x $28 + 1 x $15 = $71
2 parents and 2 children = 2 x $28 + 2 x $15 = $86
2 parents and 1 child = 2 x $28 + 3 x $15 = $101
So the $80 family pass only saves money for families with 2 or 3 children.

6 Agree
Because the number of jars of honey sold has increased each week from 8 to 23.
Or:
Can't tell for sure
Because, particularly if the market is outside, the number going to it will depend on the weather. If very few people go, then she will probably not sell many jars of honey.

7 Disagree
Ben's route is 33 cm long. The shortest route is the ⊔ shape which is 30 cm long.

8 Disagree
Because the 92 students did less than 15 minutes, homework, and that is less than half of 187 (93.5).
Or:
Can't tell for sure
Because he surveyed students on just one day. It may be that not a lot of homework is set on a Monday, or that teachers ask for it at the end of the week.

9 Disagree
Median = 16: this gives a good measure of where the centre of the data is because there are 4 values below it and 4 above.
Mean = 14.3: this has only 3 values below it and 7 above, so the median is a better measure of the centre of the data.

10 720 students voted. So, the angle should be $\frac{186}{720} \times 360°$. This is 93°, not 90°, so she is not correct.

Practice 1 (pp. 16–17)

1 **a** A   **b** 2.304 m³

   c $\frac{2}{0.1^3}$

2 **a** 38.4 minutes   **b** 11.29 am

3 **a**   **b** 20 blocks

   c 54

4 **a** One quarter of the students favoured soccer. Most students voted for rugby or soccer.

   b 42

   c Agree

     The rugby sector of the pie graph is larger than the sector for any other sport. So more students must play rugby than any other sport.

Practice 2 (pp. 18–19)

1 **a** 12, 13, 14 or 15 km

   b NE

   c 4 hours and 24 minutes

2

Letter spun	Number of spins
A	25
B	12
C	23

3 62 370 mm²

4 **a** 80 + 50 x 2 + 100

   b $69.70

   c $36.75

5 15 slices

6 $\frac{5}{11}$ or $0.4\dot{5}$

Practice 3 (pp. 20–21)

1 **a** $7.10

   b She ran to the shop, did her shopping and then walked home.

   c 1514 hours

2 **a**

   b 13 cm

3 **a** 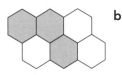   **b**

   c $130.50

4 **a** U4   **b** NW

Practice 4 (pp. 22–23)

1 **a** A bar graph

   b At least half the students did less than two hours' exercise per day.

2 **a** Rotation   **b** 24 tiles

   c 40 cm

3 **a** 064408   **b** 8 hours and 12 minutes

4 **a** 3250 mL   **b** 1500 calories

   c 24 g

Practice 5 (pp. 24–25)

1 **a** June and July

   b Either:

     Agree

     Because on average, seven months of the year have significantly less than an average of 6 hours of sunshine per day, so it's unlikely that April or October will get enough sun to lift their average over 6 hours.

     Or:

     Can't tell for sure

     Because, although seven months have less than an average of 6 hours of sunshine per day, it is possible that the average will rise significantly. We cannot be certain about what will happen in the future.

2 **a** 4800 cats   **b** $\frac{8}{24}$ or $\frac{1}{3}$ or $0.\dot{3}$

   c 1 134 000

3 **a** 7.5 cups

   b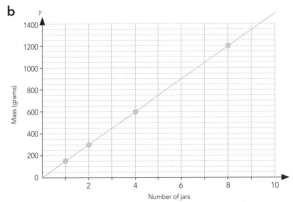

   c To make 7 jars of jam you would need 1050 g of berries.

   d $47.92

   e $2.38

Practice 6 (pp. 26–27)

1 a 4B b 55 million km
2 a D b 12%
   c 5 cups
3 a $766.06 b $1145.76
4 a $\frac{3}{14}$ or 0.21 (2 dp)
   b Two siblings
   c

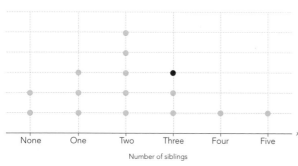

Number of siblings

Practice 7 (pp. 28–29)

1 a $8.99 for 4kg b ⌘
   c Five hundred and twenty-five thousand and thirty-one tonnes.
2 a NE b 10.36 am
3 a 6500 people b 75 huhu grubs
   c $\frac{7}{12}$ or $0.58\dot{3}$
4 5 feet 5 0.75 m

Practice 8 (pp. 30–31)

1 a Car C is the oldest and the most expensive.
   b $5850
   c 32 L
   d 23 km/hour
2 a 76% b 1 105 000
3 9 times the area of the original.
4 a If she makes four cupcakes, she will need nine chocolate buttons.
   b

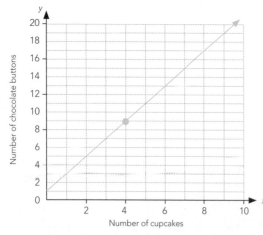

Plus any other two points along this line.

Practice 9 (pp. 32–33)

1 a $1279.22
   b Either:
      Agree because he has just bought a TV, which is a one-off big item. Otherwise he would have plenty of money in his account.
      Or:
      Disagree because he is overdrawn just two days after he has been paid.
      Or:
      Can't tell for sure because this is just four days of accounts. We should look at a lot more in order to make that judgement.
2 a The area of the walls.
   b 15 green and 5 white
   c $195.50
3 a $9 b A and C
4 a C4 b SE
   c Deer

Practice 10 (pp. 34–35)

1 a Sunday b 2.00 pm
   c NZ$600 d 19%
2 a 62 minutes or 1 hour and 2 minutes
   b (12.8 x 9.5) ÷ 2
3 a 25 seconds b 26.23 s
   c Disagree
      Because lane 6 had the second-slowest runner. Lane 4 was first, but second and third were in lanes 7 and 3.
   d 120°

Practice 11 (pp. 36–37)

1 a 372 m b 72 ÷ 2.4 + 1
   c 0.6 m
2 a 12 staff b $\frac{9}{56}$ or 0.16 (2 dp)
3 a 7.47 am b 38 minutes
   c 10°C
   d 105, 110, 115, 120 or 125
   e Agree
      Because the total height of the combined bars for 'Early' and 'On time' would be shorter than the bar for 'Late'.

Practice 12 (pp. 38–39)

1 a Overall the price of a bag of carrots is increasing.
      A bag of carrots was cheaper in week 3 than it was in week 5.
   b $27.60
2 a 4 minutes 16 seconds
   b $\frac{32}{48}$ or $\frac{2}{3}$ or $0.\dot{6}$ c 77 students

   ISBN: 9780170474474

3 a 55 km/hour   **b** SE
   c 145038   **d** $8.2 \times \dfrac{69}{100}$
   e $408

Practice 13 (pp. 40–41)

1 a 240 cm or 2.4 m
   b Rotation, reflection (both required)
2 a 0.87 m or 87 cm   **b** 80 km/hour
   c Unfair
Because the area of the big truck is four times the area of the small truck. To be fair, it should have twice the area.
3 a Point A shows that if there are 12 students, they will need 5 L of milk.
   b Need 8.5 L, so must buy 9 L. (Just 8.5 L is not a correct answer.)
   c It would become steeper.
4 a 34 students
   b Disagree
Because 11 Year 10 students want to play tennis, and 6 want to play cricket; 11 is not double 6.

Practice 14 (pp. 42–43)

1 a

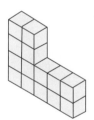

   b 10.19 pm
2 a 7 slices   **b** 36 x 36 x 5
   c $16.28
3 a 32%   **b** $33.60
   c 60°
4 a 6.8 or 6.9 cm
   b 500 000 000

Practice 15 (pp. 44–45)

1 a 9 years   **b** 5.53 pm
2 a C   **b** 2¾ cups
   c 1.125 or 1⅛ cups
3 a 4C   **b** 14 100 000
   c 5
4 a D   **b** C

Practice 16 (pp. 46–47)

1 a (36 − 23) x 15
   b 43 children
   c 1 hour and 53 minutes

2 a

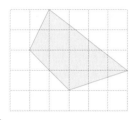

   b 1960 cm³
3 a Agree.
Because 50 liked toast and only 33 liked cereal.
   b 45 servings   **c** $4.98
   d $\dfrac{1}{12}$ cup   **e** 5.19 pm

Practice 17 (pp. 48–49)

1 a 43.8 m²
   b 1.3, 1.4, 1.5, 1.6 or 1.7 m
   c $\dfrac{22}{40}$ or $\dfrac{11}{20}$ or 0.55
   d $20.05
   e 57.5%
2 a $150 per week   **b** $450
   c Agree
Because 94 cents is just under a dollar, so the total cost should be a bit less than $45.
3 a 7°C
   b 9 hours 9 minutes

Practice 18 (pp. 50–51)

1 a NW   **b** 15 minutes
   c 098130   **d** 2.87 L
   e 460 km
2 a Bar graph   **b** 32%
   c 178 pages
3 a 6.5 x 35 + 5.5 x (35 x 1.5)
   b Runners

Practice 19 (pp. 52–53)

1 a 13.7 km   **b** 32 seconds
2 a 130, 140 or 150 mm
   b 7 minutes 30 seconds
   c 6 years and 0 months

3 a

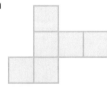

   b 64 cm³
4 a Agree
Because 3 and 4 year olds took about 16 minutes, but 7 to 9 year olds took about 3 minutes.
   b The child was 4 years old and took 3 minutes to do the jigsaw.

c

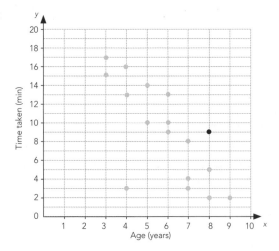

Practice 20 (pp. 54–55)

1 a

  b 3¼ or 3.25 cups
  c 1 cup flour: $0.25
  d 1¾ cups milk: $1.40
  e 2 eggs: $1.20
  f Total: $2.85

2 The mode of her marks was 16.
   The range of her marks was 8.

3 a B
  b   **c**

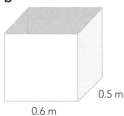 

                    0.5 m
       0.6 m

  d 6×0.2^2

Practice 21 (pp. 56–57)

1 a 244 000 steps   **b** 9632
2 a 0.3 or $\frac{12}{40}$ or $\frac{3}{10}$ **b** 55%
  c $6 \div 40 \times 360°$

3 a

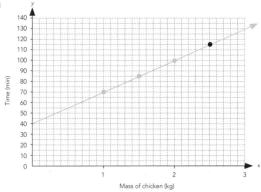

  b See graph.
  c It shows that a chicken with mass 2.5 kg will
     take 115 minutes to cook.
  d $9.89
  e 5 chickens (not 4.7 chickens)

Practice 22 (pp. 58–59)

1 a J5   **b** 92 000 000
  c 1.4%
2 a 19:26:45
  b The temperature had ~~increased~~/decreased by
     11.9 degrees.
3 a 3.6 ha   **b** 764 kg
  c 0.5 ha
4 a 21 faces   **b**

Practice 23 (pp. 60–61)

1 a Ariana and Uncle Davey.
  b Manu is taller than Uncle Davey, but lighter.
     Nick is taller and heavier than Brianna.
  c Agree
     Because it would cost $416 for separate tickets
     but just $337 for a family ticket.
2 a $\frac{1}{5}$ or 0.2   **b** $\frac{3}{5}$ or 0.6
  c 72°
  d Either: Agree
     Because he did only 200 spins and all the
     results were within 6 of 40.
     Or: Disagree
     Because $1 got 45 spins of $1 and 34 spins of $3,
     but there should have been about 40 of each.
     Or: Can't tell for sure
     Because he did only 200 spins so he might have
     got the 45 and 34 by chance.
  e Do a lot more spins.

ISBN: 9780170474474

Practice 24 (pp. 62–63)

1 **a** $162 **b** 0.864 m³

  c $\dfrac{1.2 \times 1000}{25} \times 8$

  d It had risen 8°.

  e Bar graph

2 **a** 5 minutes 20 seconds

  b 1 drip per 3 seconds = 20 drips per minute

                              = 2 mL per minute

                              = 120 mL per hour

                              = 2880 mL per day

                              = 2.88 L per day

3 **a** $2.4^3 \times 5^2$

  b

Practice 25 (pp. 64–65)

1 **a** 1, 1.1 or 1.2 km

  b NW

  c 1 hour and 24 minutes

2 **a** 3.38 pm **b** 467 000

  c 149 500 (not 149 502)

3 **a** $90 **b** 66.6̇%

4 **a**

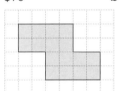

  b

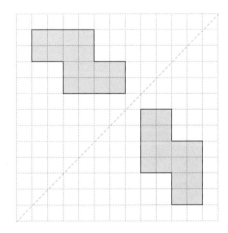

Practice 26 (pp. 66–67)

1 **a** The price of 1 kg of lemons dropped by 80 cents between weeks 3 and 5.

  b $5.40 per kg

2 **a** Bag C **b** Bag C

3 **a** 7 Days

  b $190 (or $200) + $60 + $80 = $330 (or $340) Otherwise check with your teacher.

  c $270.60 (or $246)

d 21 − (21 ÷ 7)

e 1 hour 56 minutes

f 39 minutes

Practice 27 (pp. 68–69)

1 **a** 76 years **b** 42.5%

  c Bar graph

2 **a** 8 stakes

  b More stakes than before.

3 **a**

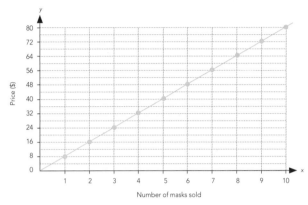

Number of masks sold

  b 63 masks

4 **a** $\dfrac{4}{5}$ **b** 1921

  c 9.6°C

Practice 28 (pp. 70–71)

1 **a** 4.164 cm **b** 10 ÷ (0.0125 × 20)

  c 5.205 mm

2 **a** 0.22

  b 115° (not 115.2°)

3 **a** ✳

  b

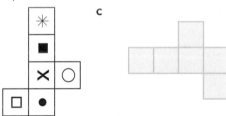

  c

4 **a** Seven days at $139 = $973. So he would have saved $74 if he had bought a season pass.

  b $245.70

Practice 29 (pp. 72–73)

1 **a** One student slept for less than two hours.

  b 17 students

  c 6.09 am

2 **a** 650 g for $5.50 **b** 30 mince savouries

3 **a** 7 **b** ✠

4 **a** $943.00 **b** 3.34 kg

  c 0.3 %

Practice 30 (pp. 74–75)

1 a $\frac{1}{15}$ or $0.0\dot{6}$ **b** $\frac{12}{15} = 0.8$

2 a A   **b** 8 cm

c

3 a 6130 (not 6131 or 6131.04)
  b 160 911
  c Agree
  Because compared with the New Zealand average, there are more people in Hamilton that are in the 0–14 age group and fewer in the 65+ age group.

4 a 320 kg   **b** 3 hours 15 minutes

Practice 31 (pp. 76–77)

1 a Scatter graph and line graph.
  b 4 kg
  c

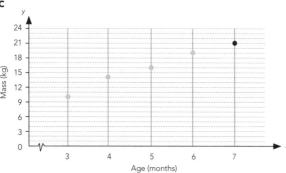

  d 20 g/kg food will be 600 g per day
  One bag = 15 000 g
  $\frac{15\,000}{600} = 25$ days

2 a 6 games   **b** 150°
  c 1.75 cups   **d** $99.68
  e 3.1° or 3.2°   **f** Every 6 hours

Practice 32 (pp. 78–79)

1 a 6.16 m²
  b 1.8, 1.9, 2.0, 2.1, 2.2 or 2.3 m
  c $6^2 + \frac{1}{2}(3 \times 6)$
  d 1.1 kg

2 a Less than a quarter like Vegemite.
  b 8 students
  c Agree
  Because the sector for Vegemite is a little smaller than the sector for jam, which has an angle of 90°.

3 a Six centuries
  b 15.3 kg
  c

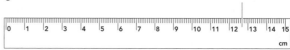

Practice 33 (pp. 80–81)

1 a Day 3
  b North Island brown kiwi

2 a $1019.15   **b** $690.00
  c 91 psi

3 a 600 pumps
  b 1 in 1000
  c No. He did 60 + (1.8 x 32). He should have done (60 x 1.8) + 32 = 140°C

4 a $27.5 x 1.5 x 6 = $247.50
  b 10.5%

Practice 34 (pp. 82–83)

1 a

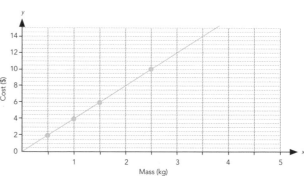

  b See graph.
  c 2.5 kg of kūmara costs $10.
  8 or 9 plants.
  Possible reasoning:

  d $2.24
  e 325 g

2 a 0.7 m at $5.60 = $3.92
  0.6 m at $7.50 = $4.50
  So fabric A is cheaper by 58 cents.
  b B
  c Mean and median.

3 a 12 seconds   **b** 25 kg

Practice 35 (pp. 84–85)

1 a 14:26   **b** 29 minutes
  c 1.42 pm

2 a Mass of trays and kiwifruit ≈ 250 trays x 4 kg
= 1000 kg
Total with pallet mass ≈ 1050 kg
Check with your teacher if you have a different method.

b 28 pallets (not 28.8 or 29)

3 a 2019

b Agree
Because in every year there have been more babies called Maia than Aurora.
Or:
Can't tell for sure
Because the popularity of Aurora is increasing more rapidly than the popularity of Maia.
Over the three years the number of babies called Aurora has increased by 19, but Maia has only increased by 8.

c 0.31%

d 12 400

e 5 083 426 or 4 506 459

Practice 36 (pp. 86–87)

1 a 0.45

b Agree
Because the percentage of highly physically active and physically active people has increased in 2020–2021, while the percentages of those doing insufficient physical activity or little or no physical activity has decreased.
Or:
Disagree
Because while the percentage of highly physically active and physically active people has increased in 2020–2021 and the percentages of those doing insufficient physical activity or little or no physical activity has decreased, none of these changes is by more than 1%.

2 a C

b B

c Disagree
Because what results he had in the past will have no effect on what happens next. The probability of getting white on any throw will always be a third.

3 a $126.75

b $118.93

c $118.00

d

e One gram of coffee costs $\frac{880 \text{ cents}}{200} = 4.4$ cents
So 35 g coffee costs 35 x 4.4 cents = 154 cents
= $1.54.
Check with your teacher if you have a different method.

Practice 37 (pp. 88–89)

1 a At least 3 times.

b

c 220 000 000

d 45, 50 or 55 cm

2 a 34%

b 1 kg at the supermarket costs $\frac{\$17.95}{5} = \$3.59$.
They were $3.59 – $2.80 = 79 cents per kilogram cheaper at Matiu's.

3 a 55.44 cm² **b** 60°

4 a 0°C

b Disagree
Because there was a frost on only nine mornings which is less than a third of 31 (10.$\dot{3}$).

Practice 38 (pp. 90–91)

1 a 62.8 km **b** 195.0 km

c 68 km/hour **d** NW

e $7.00

2 a

b Slightly thinner than the recipe says.

c 01:12 or 1.12 pm

d 24 pieces

e $1.38

1 a

b 24   **c** 17 blocks

2 a 16.7%   **b** 4 160 000

3 a 7.5 hours

b 100 m in 2 minutes = 0.1 km in $\dfrac{2}{60}$ hours

$= 0.1 \times \dfrac{60}{2}$ km in 1 hour

= 3 km/hour

c 5 rotations

d 27 km/hour

e NW

1 a 15:24

b Gurnard

c Disagree

Because they didn't catch the largest number of fish (13) on the second day. They caught 14 fish on day 3.

Or:

Can't tell for sure

Because we are told only the numbers of fish. They may have caught really big fish on another day.

d 3 000 000

e mean = median > mode

2 a 54 cm²

b

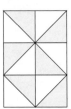

3 a If the short sides of the triangle were equal, then the angle would be 45°.

However, the side opposite the angle is longer (0.95 cm), which means the angle must be a bit more than 45°, so 50° is a good estimate.

b 9 buckets

c 40 kg bags cost 40.2 cents per kg
20 kg bags cost 41.0 cents per kg
15 kg bags cost 40.8 cents per kg
So the 40 kg bag gives him the lowest price.